Knowing Global Environments

STUDIES IN MODERN SCIENCE, TECHNOLOGY, AND THE ENVIRONMENT

Edited by Mark A. Largent

The increasing importance of science over the past 150 years—and with it the increasing social, political, and economic authority vested in scientists and engineers—established both scientific research and technological innovations as vital components of modern culture. Studies in Modern Science, Technology, and the Environment is a collection of books that focuses on humanistic and social science inquiries into the social and political implications of science and technology and their impacts on communities, environments, and cultural movements worldwide.

Mark R. Finlay, *Growing American Rubber: Strategic Plants and the Politics of National Security*

Gordon Patterson, *The Mosquito Crusades: A History of the American Anti-Mosquito Movement from the Reed Commission to the First Earth Day*

Jeremy Vetter, ed., *Knowing Global Environments: New Historical Perspectives on the Field Sciences*

Knowing Global Environments

New Historical Perspectives on the Field Sciences

Edited by

Jeremy Vetter

10/2/10

For Ann,

With thanks for all you have done for me.

With best wishes,

[illegible]

Rutgers University Press

New Brunswick, New Jersey, and London

Library of Congress Cataloging-in-Publication Data

Knowing global environments : new historical perspectives on the field sciences / edited by Jeremy Vetter.
p. cm.
Includes bibliographical references and index.
ISBN 978-0-8135-4875-3 (hardcover : alk. paper)
1. Science—Fieldwork—History. I. Vetter, Jeremy, 1975–
Q175.K557 2010
507.2′3—dc22

2009052308

A British Cataloging-in-Publication record for this book is available from the British Library.

Visit our Web site: http://rutgerspress.rutgers.edu

Manufactured in the United States of America

Contents

Figures and Tables

Figures

Table

Acknowledgments

This book originated from a workshop conference we organized at the University of Pennsylvania on May 10–12, 2007, under the same title, held in honor of Robert E. Kohler. To honor Rob's legacy as a scholar who has always been eager to engage with new and vital areas of historical inquiry, we designed the workshop from the beginning not as a traditional festschrift but instead as a gathering of scholars on a theme of broad importance to both historical scholarship and the public at large: how field scientists in a wide variety of disciplines have produced knowledge beyond the local level. Participants were selected not necessarily because of their direct connection to Rob as former students or colleagues (some were, others were not), but because of their engagement in exciting research at the forefront of the history of the field sciences. Papers were pre-circulated, and the result was a lively and engaging workshop that led to further rounds of revision and, ultimately, the book you now hold in your hands.

I am pleased to acknowledge the other co-organizer for the workshop conference, Susan Lindee of the University of Pennsylvania, as well as the institutional sponsors: Philadelphia Area Center for History of Science, Academy of Natural Sciences, American Philosophical Society, Chemical Heritage Foundation, Princeton University, and the University Research Foundation and Department of History and Sociology of Science at the University of Pennsylvania. Special thanks are due to the other scholars who participated in the workshop as presenters or commentators and whose contributions were crucial to the discussions that shaped the essays in this volume: Michael Bravo, Emily Brock, Eve Buckley, Graham Burnett, Alex Checkovich,

Tom Gieryn, Drew Isenberg, Christine Keiner, Scott Kirsch, Henrika Kuklick, Naomi Oreskes, and Phil Pauly. Appreciation is also due to the many other attendees at the workshop for their ideas and questions. We are also grateful to all those in the Department of History and Sociology of Science who helped make the workshop a success, including the administrative staff, students, and faculty. To Rob, whose works and ideas have been so important to so many scholars, we express our enduring gratitude. Finally, thank you to acquisitions editor Doreen Valentine, series editor Mark A. Largent, and the anonymous external reviewer for Rutgers University Press for their valuable suggestions that helped to improve this book.

Knowing Global Environments

Jeremy Vetter

Introduction

We live in an era of great concern about environmental sustainability. Global problems such as climate change, resource depletion, and biodiversity reduction have worked their way into both public and scholarly consciousness with a force not felt in at least a generation. Public awareness of these and other environmental issues in the modern world has often rested on scientific knowledge and research practices. This is especially the case for problems that transcend the local level; everyday experiences have often proved insufficient to comprehend larger-scale environmental phenomena. Moreover, modern scientific knowledge has become centrally important for solving or mitigating these environmental problems. While citizens and policy makers who are interested in addressing environmental issues may differ in their estimation of how vital science may be—ranging from those who regard the discovery and application of scientific knowledge as the primary tasks of earth repair to those who believe that changes in values, social and economic systems, and institutions are more important—few would dispute the centrality of the modern environmental sciences for influencing how those problems have been represented and understood.

Correspondingly, awareness of the importance of the environmental sciences in the field among historians of science has been rising in recent years.[1] This book brings together original historical work analyzing a wide range of field sciences broadly concerned with the environment, including archaeology, agricultural science, botany, climatology, ecology, evolutionary biology, oceanography, ornithology, and tidology. Studies of local place and practice have already demonstrated the importance of the field site to how environments have become known in particular places.[2] Now scholars and citizens are becoming increasingly interested in the production and circulation of knowledge at a variety of scales beyond the local, including the region, nation, empire, and globe. This agenda points to the wider significance of fieldwork for the history of science, and indeed for the history of the modern world more generally. By providing historical perspective on the emergence of ideas and practices in the field sciences, this book aims to identify and analyze key issues for historians of science, environmental historians, general historians, scholars in related disciplines, field scientists, and members of the broader public who care about knowing global environments in modern society.

But what, exactly, constitutes "the field"? In its broadest meaning, the field can be anywhere outside the laboratory where scientists have worked. While laboratories have aimed to produce universal, placeless knowledge, field sites have produced knowledge that is based in place.[3] Historically, as rhetorically placeless laboratories ascended to their high epistemic status in modern science by the late nineteenth century, "the field" was simultaneously reconstructed as the residuum of messy, complex, and uncontrollable nature. In the wake of this shift, many field scientists of the early twentieth century strove to bring elements of the laboratory into the field, including its instruments, its experimental protocols, and its quantitative precision.[4] Others chose to assert the value of place-based research in the natural world as distinguished from the artificial environment of the laboratory. While unabashedly field-based research never achieved anything like the funding and prestige of laboratory science during the twentieth century, neither did it disappear or diminish—rather, it proliferated—in its absolute frequency.[5]

Scaling Up

Not only has fieldwork proliferated in this age of laboratories, but field scientists have found their own distinctive ways to pursue a more general, global, and universal knowledge. Rather than constructing a "placeless place" for research practice in which knowledge would be ready-made as universal, as laboratory scientists did,[6] field scientists have tried to achieve ever greater scope for the knowledge they have produced. They have extended place geographically by moving from the local to higher levels of territorial scale. Along the way, they also opened up new spaces for science, such as oceans and polar regions.[7] In other words, field scientists have not always remained content to investigate the particularities of local places—though they have certainly done that sometimes, and have often come to identify with and remain fiercely loyal to their local field sites—but they have also found ways to make knowledge claims at higher scales. Robert E. Kohler's concluding essay ("History of Field

Science") calls this approach the "long degree," in an insightful spatial analogy with the temporal *longue durée* of Fernand Braudel.

In scaling up, the region has been a powerful analytical unit for considering how field scientists have extended the value and significance of fieldwork. At first glance, the region may seem a modest scale for extending claims from local field sites, particularly when that region is not a broad world macro-region such as the global tropical rain forest belt or the Eurasian steppe, but instead an environmental region wholly within one country. Yet by defining and studying an environmental region within which the natural world was thought to be uniform enough to merit robust generalization, field scientists have extended the significance of their research findings through geographical breadth. One obvious strategy for producing regional knowledge has thus involved comparing and aggregating data from many localities, thereby fending off the criticism that one place cannot be made to stand for an entire region.[8] Defining the region has itself been a significant move, since claims to extend knowledge beyond the local can thus become hopelessly entangled in debates about the proper boundaries of the region being scaled up, disputes that have been at least as much political or cultural as scientific. But engaging in such debates has been a necessary occupational hazard for field scientists who wish to expand the purview—and thus, both the scientific prestige and the practical applicability—of their findings.

But extending knowledge to larger areas by aggregating multiple sites is not the only way that regions have been important in science. In some cases, regions have been especially valued for their environmental diversity. Consider mountainous areas, with their steep slopes. As Jeremy Vetter's chapter ("Rocky Mountain High Science") suggests, the mountains became an appealing place for combining the advantages of long-term residence during summer vacation at a field station with access to these diverse environmental types. Likewise, the second half of Michael Reidy's chapter ("From the Oceans to the Mountains") analyzes the importance of mountain slopes to the fieldwork of Alexander von Humboldt and Joseph Hooker. On the other hand, the chapters by J. Conor Burns ("Negotiating the Agricultural Frontier in Nineteenth-Century Southern Ohio Archaeology") and Mark Barrow Jr. ("On the Trail of the Ivory-Bill") show how the field sciences of archaeology and ornithology flourished in two different U.S. regional contexts—the farming frontier of the Ohio Valley and the forests of the Southeast. In these cases, it was not so much the environmental diversity of the regions that mattered as their richness for highly valued scientific objects (artifacts or rare birds). By pursuing a strategy emphasizing the special characteristics of a region—richness or diversity—and thereby making it especially suitable for making knowledge claims, researchers could achieve greater significance for their findings. In other words, just as local field sites could be unique, significant, or otherwise powerful enough to merit cosmopolitan interest, so too could entire regions.

Beyond the region are a wide variety of larger geographical units of analysis, such as empires, oceans, and the entire globe. As noted above, the epistemic power of the vertical geography of mountains is evident not only in Vetter's essay deploying a regional scale of analysis as noted above, but is also examined in its larger imperial context in Reidy's chapter. Reidy productively compares the verticality of field science in the mountains with the global horizontal extension of tidology—the study of the tides—across the oceans. Indeed, as places between the land-based territories where field scientists normally live yet which occupy the vast majority of the Earth's surface, oceans have been key sites for the development of field sciences over a large geographical scale. "Knowing the oceans" as large-scale natural phenomena could also involve learned elites with the practical knowledge of mariners, as Joyce Chaplin's recent study of the Gulf Stream indicates.[9] Moreover, as Helen Rozwadowski ("Playing by—and on and under—the Sea") argues in her chapter, oceans became places that were increasingly known through practices that intertwined work and play.

Both Reidy and Rozwadowski suggest the importance of global travel to scientific fieldwork, but in Lynn Nyhart's chapter ("Emigrants and Pioneers") movement across geographical space takes center stage. Nyhart explores the rich intersection of thinking about human migration and the development of evolutionary theorizing in nineteenth-century Germany, as exemplified in Moritz Wagner's attempt to put geographical isolation through migration to new environments at the center of evolutionary thought. Long-distance movement and evolutionary change are also central themes for Stuart McCook ("Managing Monocultures"). He examines agricultural field sciences, including not only movement in space through the global exchange of coffee varieties but also the challenge posed by evolution as cultivated organisms and their pathogens changed ontologically. Scientists thus confronted what Edmund Russell has called "evolutionary history."[10]

Finally, field scientists in recent years have come to focus much of their time and energy on what is increasingly recognized as perhaps the most critical and visible global environmental knowledge domain of our times: the shared global atmosphere. James Fleming's chapter ("Planetary-Scale Fieldwork") considers how scientists constructed climate knowledge that literally encompassed the entire globe as its unit of analysis.[11] Knowledge production at such large geographical scales has depended on the "infrastructural globalism" identified by Paul Edwards in his analysis of observation networks such as the World Meteorological Organization's World Weather Watch in the 1960s.[12] Thus, from the region to the globe, these essays provide a range of diverse perspectives on the extension of field science beyond the local.

Making Environmental History

Another central goal of this volume is to demonstrate the relevance of the history of the field sciences to environmental history. This increasingly close relationship has operated in several key ways, all of which are demonstrated in analytically suggestive case studies in this volume. Most obviously, from the regional to the global scales, work in the field sciences has related directly to pressing environmental domains, such as biodiversity, oceans, and the atmosphere. As Barrow's chapter indicates for the case of the ivory-billed woodpecker, field scientists have debated and defined both the existence and geographical distribution of endangered species. Without a field science operating on at least a regional scale, it would be impossible to conceptualize such problems as the threat to biodiversity posed by economic development over the entire ranges of various animal species.[13] The same might also be said about the studies in ecological botany conducted at the Rocky Mountain field stations discussed in Vetter's chapter, which aimed to understand the dynamics of plant formations in diverse environments along a mountain slope. Similarly, in the case of the ocean scientists examined in Rozwadowski's chapter, field practices were connected with the spatial elaboration of knowledge claims about this vast and potentially fragile domain. Finally, the relevance of Fleming's chapter to understanding global environmental problems is self-evident, since it analyzes the historical production of knowledge about the causes of and potentially dangerous responses to climate change. Given how closely intertwined human understanding of these looming environmental problems have been with forms of scientific knowledge constructed in the field at varying scales, the history of the field sciences is an essential component of these problems' environmental history.

Even in the other chapters, where the science in question is not necessarily directly related to the construction of knowledge underlying our understanding of a current environmental crisis, there are important and rich linkages with environmental issues that have been important historically. In the past, the field sciences have been at least as important for development and exploitation of the environment (see below) as for environmental concern. The conceptualization of the entire globe as an arena for imperial expansion, as explored in Reidy's chapter, offers one reminder of the role of the field sciences. Likewise, on a more circumscribed scale, the mounds uncovered during the agricultural development of the Ohio Valley, as explained by Burns, revealed to nineteenth-century archaeologists a landscape formerly developed, depopulated, and then developed again, thus providing clues to an important linkage in how they—and, not incidentally, the environmental historians who draw on the work of those archaeologists and their successors—came to understand interactions between humans and the environment over the long term. Connections with environmental transformation are also visible in other chapters, including Nyhart's consideration of the role of migration in processes of human-environment interaction and McCook's intertwined local and transnational perspectives on the role of knowledge in the agroecologies of global coffee production.

Not only did the field sciences integrally shape the understanding of environmental problems, but particular environments in turn shaped the development of those very same field sciences. Such influences are perhaps most obvious at the regional level. The dense forests of the U.S. Southeast, for example, where Barrow's field ornithologists searched for the ivory-bill, presented a very different set of experiential constraints than the cleared farmlands of the Ohio Valley explored by Burns's archaeologists. Likewise, the open meadows of the high mountain valleys of the Rockies, where the Rocky Mountain field stations Vetter discusses were located, provided a very different sort of place to practice science from the marine environments investigated by Rozwadowski's ocean scientists. What may seem like mundane issues such as daily provisioning and movement of a field party—whether in farm or forest, mountaintop or ocean bottom—demonstrate not just how the field sciences have evolved in some kind of primordial relationship to the natural environment itself, but also how particular places have been situated within networks of transportation and communication into and out of them. But the relevance of environmental context to field practice is also evident in the construction of knowledge on a larger scale. As discussed by Fleming, conducting field research for a climate scientist meant collecting atmospheric data at a variety of interlinked observation stations in order to construct global models. These conditions for scientific practice were quite different from McCook's agricultural scientists working at specific sites of tropical monocrop coffee production. The conditions of field practice, no less than the ideas produced, are thus a crucial aspect of the relationship between environmental history and the history of the field sciences.

Yet the relevance of the histories of the field sciences elaborated here extends well beyond these direct content connections of the knowledge constructed to environmental problems and the reciprocal shaping influence of environmental and human geography on practice. Indeed, it could be argued that the "environment" itself as an object of knowledge has been constituted through the field sciences. As Reidy points out in his chapter, the global construction of the environment has involved both horizontal and vertical dimensions. The horizontal geography of science—and therefore of the global environment as constructed by it—has received the most attention so far, both in this volume and elsewhere. Mapping, especially—whether of the tides examined by Reidy himself or of the seemingly innumerable other natural features such as plants, animals, minerals, topography, temperatures, rainfall patterns, or even putative human racial types—has proved to be an important means by which field sciences have constructed the environment on an increasingly global scale.[14] This horizontal geography has also extended to migrations, exchanges, and measurements around the world, such as the environments constructed by field scientists explored in the chapters by Nyhart, McCook, and Fleming.

The vertical dimension of constructing the environment has also been crucial and is now receiving increased attention. In this volume, Reidy's mountaineering botanists, Burns's underground archaeological resources, Rozwadowski's ocean scientists, and Vetter's mountain field stations illustrate some of the diverse ways that field scientists have constructed the environment as extending upward or downward.[15] This verticality, as Bruce Braun has shown in the case of geology, represents far more than the simple uncovering of previously hidden knowledge, involving the active construction of the underground through the reiterative engagement of field scientists bringing new epistemic frameworks to the field. The same could be said of the aboveground vertical dimension evident in attempts by climate scientists to model the atmosphere's interactions with the land below. The verticality of the environment, no less than its horizontality, is produced through the practices of scientific fieldwork.

Field Developments

This environmental history of scientific fieldwork has been integrally connected with global processes of imperial expansion and capitalist development. The uses of knowledge for environmental exploitation (and conservation) have constituted the political and economic framework for connecting the history of the field sciences with the histories of natural environments. How, then, has the political economy shaped those practices and places? If the laboratory is a place that has been rhetorically separated from (while simultaneously embedded in) a material economy in order to produce fundamental knowledge that can later be adapted to practical or industrial use, the field has more often been deeply immersed from the start in a world of capitalist economic development and environmental resource exploitation.[16] Sometimes the field sciences have been directly involved in furthering development, while at other times they have responded more to a conservation- or preservation-driven agenda (and sometimes a mixture of both).[17] But almost always the field has been a "working landscape," as discussed by several of the authors in this volume and in the concluding essay by Kohler.

Consider the example of agriculture, surely one of humankind's most enduring means of transforming the natural environment. For some field scientists, such as the archaeologists examined by Burns, agriculture could be the catalyst for (and threat to) the uncovering of vital scientific resources. Even as the sod was turned to create new, more heavily manipulated agroecosystems for the development of the Ohio Valley, material evidence of past human inhabitation became the object of intense scientific scrutiny by museum-based archaeologists in Boston and Washington, D.C. While the archaeological findings may seem to be a curious by-product, this context of agricultural development, as Burns shows, was central to understanding the practice of fieldwork. By contrast, the field scientists analyzed by McCook took up agricultural development in working landscapes as their main goal. The role of knowledge production in the global exchange and expansion of plant varieties surely constitutes one of the greatest economic development legacies of practical science for transforming landscapes, from the tropical coffee plantations discussed by McCook to countless other monocultures of wheat, corn, rice, sugarcane, bananas, cotton, fruits, and vegetables that have increasingly dominated the modern world's food production system.

Similar points could be made about other landscapes and seascapes around the world, from forests and fisheries to oceans and high mountains. As Barrow demonstrates in his chapter, the logging of the forests, which destroyed habitat for rare birds, heavily influenced ornithology in the U.S. Southeast. The increasing recognition of the costs of economic development and the commodification of forest resources for wildlife habitats has indeed become a major preoccupation of field scientists concerned with biodiversity in the modern world.[18] As Christine Keiner has shown in her recent book on the "working landscape" of oystermen and biologists in Chesapeake Bay, field scientists have struggled for control over valuable environmental zones.[19]

Other environmental resource uses suggest a more benign, but no less consequential, relationship between economic development and scientific fieldwork. For example, as Rozwadowski shows in her essay, the oceans were more than just working landscapes: they were also landscapes of play. Similarly, the high mountain meadows of the Rockies, as Vetter discusses in his chapter, provided a setting for recreational and tourism development that shaped the emergence of summer field stations. Both in providing a railroad-accessible village with a built environment on which to set up the field stations and in motivating the choice of such a place to spend the summer vacation "working at play" (to use Cindy Aron's apt term), such field stations were deeply immersed in the tourism and recreational development of the Rocky Mountains.[20] And as Reidy's work here and elsewhere argues, the field science of the mountaintop cannot be divorced from the sport of mountaineering, even if the relative roles of those two activities have been shifting over time. The affective dimensions of recreation, leisure, and human adventure cannot be ignored in the history of the field sciences.

The economic development of the world has also been closely related to the processes of colonization and human migration. As Nyhart shows, the transformation of the world by German settlers provided the context for evolutionary theorizing about the roles of migration and environmental change. Moreover, as Reidy argues, the extension of more formal large-scale global empires, such as the British Empire, has been central to certain field sciences such as tidology and botany in shaping how the aims of scientific knowledge production have been defined. This work nicely complements existing histories that stress how imperial thinking shaped the production of knowledge, as in the case of geology.[21]

Environmental resource exploitation, whether in the more conventional material form of agriculture and forestry or in the more symbolic form of tourism and recreation (which have often concealed their own material transformations), has fostered the growth of a set of field sciences that have been part and parcel of the global development apparatus. Yet there have also been subversive field sciences, such as ecology, that at least in part have sought to provide alternative forms of knowledge for use.[22] Indeed, the field sciences in general hold out at least the possibility of a more holistic and contextualized form of scientific knowledge that can recognize the environmental and human effects of economic development. We have certainly experienced a recent surge of interest in knowing global environments to calculate, quantify, and measure biodiversity loss, soil degradation, resource shortages, industrial pollution, and—perhaps most important of all—human-induced climate change.

But the hope of the field sciences to produce credible knowledge to address these environmental crises should also be tempered by an awareness of how deeply our structures of knowledge and beliefs about how to act on that knowledge are embedded in a modernist developmental model.[23] As Fleming points out in his chapter, schemes to devise a "technological fix" to climate change exude a disconcerting hubris and seem to ignore the cautionary tales stemming from a realistic historical assessment of past attempts to engineer the atmosphere. Regimes adopting both capitalist and communist ideologies have fallen prey to such hubris in other domains of environmental control. Whether in the terms offered by the essays in the volume, or in the guise of other scholarly frameworks such as James Scott's "high modernism," Paul Josephson's "brute-force technologies," and countless other variations, historians and social scientists have increasingly recognized the problems inherent in our structures of knowledge for development following both American and Soviet models.[24]

Public Participation

Paralleling the rise of interest in environmental and sustainability problems since the 1970s has been a growing enthusiasm for public participation in science, which might in some ways appear to mitigate the top-down bias in modern structures of scientific knowledge. Indeed, the involvement of lay people in science, whether they are conceived of as participants in the process of knowledge creation or simply as a public audience for it, has become something of an established trend in the social studies of expertise.[25] Of course, broader involvement in scientific knowledge production and interpretation need not imply a reduction of status differentials; the subversive status of public participation in science has often been uncertain and contested.

As the cases in this volume show, such dilemmas and conflicts over the involvement of lay people in the field sciences have a long history. Perhaps the most obvious way that lay people have been involved in scientific fieldwork has been through their direct participation as field assistants. In his analysis of efforts to locate and the study the ivory-bill, Barrow's chapter highlights the role of the field assistant as a crucial part of the practice of fieldwork. Likewise, Joseph Hooker's mountaineering expeditions discussed in Reidy's chapter would have been impossible without the contributions of local guides and porters. Wherever one looks in the field sciences, the "technicians of the field" have been there.[26] Field assistants have been ubiquitous and essential in the field sciences, and hardly any expedition, survey, station, or other enterprise could be effective without them. Subordinate, yet valued for their skills, field assistants constitute a social category worthy of further investigation in the history of scientific practice.

More broadly, members of the public have been involved in fieldwork not just as hired field assistants but also simply by virtue of their residence in or near field sites. As Burns shows in his chapter on Ohio Valley field archaeology, settler farmers were crucial participants in the networks built by eastern museum archaeologists, uncovering the mound structures and artifacts that were so crucial to their research and controlling the land on which they were found. Negotiating and working with people who live there has thus been a crucial part of doing science in the field. This was true of both land-based fieldwork, with its more obvious local residents who dwelled in the field, and sea-based fieldwork, which brought together ocean scientists with the personnel of the long-distance sea voyage, as Rozwadowski has shown in a previous essay.[27] Of course, while being on the home turf (or ship deck) of lay people who lived there might have provided some leverage against the conventional hierarchy of expertise, it rarely overturned it.[28]

The durability of hierarchical structures of expertise in the field sciences can perhaps best be seen in the outcomes of attempts to transcend those emerging status categories in the mid-nineteenth century. As Nyhart's chapter in this volume and McCook's earlier work on Paul du Chaillu both indicate, those judged to lack professional expertise, yet who tried to claim the role of offering larger interpretations of the observational facts they had gathered in the field, confronted clear barriers.[29] The contest over the qualifications of the field observer for epistemic status may never be fully settled, but it is clearly a boundary line that has been policed often by credentialed experts over at least the past century and a half.

The boundary line between lay and expert has also been crossed innumerable times, most obviously when members of the public—as formal or informal students of science—are transformed into experts. The process of field education has involved many pathways since the nineteenth century, but one general shift has been away from informal field apprenticeship to a more systematic training in field practices embedded within the system of higher education, from the bachelor's degree to the Ph.D. As Kohler has shown in his recent study of field zoologist Vernon Bailey's informal apprenticeship in the late nineteenth century, such an alternative form of education for the field sciences still existed just over a century ago.[30] Yet as the twentieth century dawned, the integration of field training with formal university education became increasingly the norm, even in the most outdoor-based of the sciences. As Vetter's chapter on Rocky Mountain field stations shows, the teaching function of these field sites was at the very heart of their justification and everyday operations. Only by teaching rigorous field methods both in the outdoors and yet connected to the university curriculum as a set of advanced course work, field station proponents believed, could the next generation of field scientists be properly trained. This outdoor education structured by the rigors of the university extended even beyond their own advanced undergraduates (and, in a few cases, graduate students) to include continuing education for school teachers as well.

Yet however they were educated, whether through informal apprenticeship or (increasingly) through formal university courses and mentored research delivered in the field, the students of the field sciences found themselves crossing an important epistemic line. In being taught to perceive environments from the regional to global scales through scientific categories, instruments, and methods, they were coming to adopt a new perspective on the natural world they encountered—what Kohler has instructively called the "cosmopolitan" way of knowing in order to draw attention to how it complements the "residential" way of knowing employed by people who live in a particular place.[31] By setting up the categories of cosmopolitan and residential, as opposed to some more conventional antinomy such as global versus local that suggests geographical scalability and encompassment, Kohler draws our attention to how each way of knowing the environment has offered its own insights that cannot be fully captured by the other. In one common scenario, it may be the case that while a resident of a given place knows where to find a particular object of interest in a given season, a cosmopolitan field scientist is more likely to know how to classify that object in a larger scientific taxonomy. In the field of vertebrate paleontology in the early twentieth-century American West, for example, resident ranchers might be valued for their knowledge of where to prospect for new fossil deposits, while visiting scientists had greater authority in the ultimate categorization of the specimens thereby discovered.[32] Only by fully recognizing and taking seriously both ways of knowing—residential *and* cosmopolitan—can historians of the field sciences understand how the field sciences have developed and how they have operated in diverse sites of work and play.

Indeed, the contrast between residential and cosmopolitan knowledge can offer a fitting conclusion to this introductory essay, for the distinction between the two highlights a key challenge that has confronted field scientists who have attempted to produce knowledge beyond the local. Since modern science in general has valorized cosmopolitan knowledge, and has only accorded merit to forms of residential knowledge—constructed as nested in the "indigenous" or "local"—that could be subsumed within it,[33] field scientists moving to larger scales from regional to global have to some extent faced the same pressure. In practice, residential knowledge rarely seems to scale up easily, at least not within the standards of rigor that predominate in modern science. Thus, as the field sciences have increasingly embraced global or cosmopolitan standards, they have run the risk of losing their deep engagement with particular places and the understandings of the natural world of people who live, work, and play in them.[34] Can members of the public at large be convinced to care deeply about environmental problems, such as climate change, whose comprehension requires transcending everyday local experience? Despite their historical shaping by modernist developmental frameworks, one of the key possibilities for the field sciences in the present moment may be how they could also become a means for striking a balance between the globalizing and universalizing ambitions of modern knowledge-making and the practical needs of a world of tremendously complex and variable environments whose knowing matters so much for human survival and sustainability.

Notes

1. Likewise, since the publication of the 1996 *Osiris* volume titled *Science in the Field*, innovative and pathbreaking historical scholarship on the field and environmental sciences has proliferated. For a synopsis of the state of the art in the mid-1990s, see Henrika Kuklick and Robert E. Kohler, "Introduction," *Science in the Field, Osiris* 11 (1996): 1–14. For an overview of more recent work, see Kohler's essay in this volume.
2. Some insightful studies stressing the influence of local place on field ecology are Eugene Cittadino, "A 'Marvelous Cosmopolitan Preserve': The Dunes, Chicago, and the Dynamic Ecology of Henry Cowles," *Perspectives in Science* 1 (1993): 520–559; Matthew W. Klingle, "Plying Atomic Waters: Lauren Donaldson and the 'Fern Lake Concept' of Fisheries Management," *Journal of the History of Biology* 31 (1998): 1–32; Chunglin Kwa, "Modeling the Grasslands," *Historical Studies in the Physical and Biological Sciences* 24 (1993): 125–155; Daniel W. Schneider, "Local Knowledge, Environmental Politics, and the Founding of Ecology in the United States: Stephen Forbes and 'The Lake as Microcosm' (1887)," *Isis* 91 (2000): 681–705; and Christian C. Young, "Defining the Range: The Development of Carrying Capacity in Management Practice," *Journal of the History of Biology* 31 (1998): 61–83.
3. Robert E. Kohler, "Place and Practice in Field Biology," *History of Science* 40 (2002): 189–210. A thoughtful case study in the social sciences is Thomas F. Gieryn, "City as Truth-Spot: Laboratories and Field-Sites in Urban Studies," *Social Studies of Science* 36 (2006): 5–38. For a recent overview of lab history, see Kohler, "Lab History: Reflections," *Isis* 99 (2008): 761–768, along with other articles in the same special focus section by Ursula Klein, Graeme Gooday, and Thomas F. Gieryn.
4. Robert E. Kohler, *Landscapes and Labscapes: Exploring the Lab-Field Border in Biology* (Chicago: University of Chicago Press, 2002).
5. On the continuing vitality of the fieldwork tradition, especially in cultural anthropology, see James Clifford, "Spatial Practices: Fieldwork, Travel, and the Disciplining of Anthropology," in *Anthropological Locations: Boundaries and Grounds of a Field Science*, ed. Akhil Gupta and James Ferguson (Berkeley and Los Angeles: University of California Press, 1997), 185–222; and Henrika Kuklick, "After Ishmael: The Fieldwork Tradition and Its Future," in Gupta and Ferguson, *Anthropological Locations*, 47–65. For a series of practitioners' reflections on fieldwork from the mid to late twentieth century, which makes clear the perception of field science as the impoverished and underappreciated sibling of laboratory science (especially molecular biology), see Elizabeth Higgins Gladfelter, *Agassiz's Legacy: Scientists' Reflections on the Value of Field Experience* (Oxford: Oxford University Press, 2002).
6. As Kohler observes, the strategy of placelessness "can fail when people begin to think that the work of labs says more about phenomena that happen only there." Kohler, "Labscapes: Naturalizing the Lab," *History of Science* 40 (2002): 474. On the

ecology of the lab itself, see Kohler, *Lords of the Fly: Drosophila Genetics and the Experimental Life* (Chicago: University of Chicago Press, 1994).

7. On the oceans, see below. On the Arctic, see Michael Bravo, "Geographies of Exploration and Improvement: William Scoresby and Arctic Whaling, 1782–1822," *Journal of Historical Geography* 32 (2006): 512–538; and Michael F. Robinson, *The Coldest Crucible: Arctic Exploration and American Culture* (Chicago: University of Chicago Press, 2006).
8. One can find such an impulse to develop regional knowledge, for example, in the effort by the U.S. Department of Agriculture's Office of Dry-Land Agriculture Investigations to establish field stations scattered across the Great Plains region in the early twentieth century. Examples of regionalizing knowledge in other parts of the world can be found elsewhere in the notes to this introductory essay and in Simon Naylor and G. A. Jones, "Writing Orderly Geographies of Distant Places: The Regional Survey and Latin America," *Ecumene* 4 (1997): 273–299.
9. Joyce E. Chaplin, "Knowing the Ocean: Benjamin Franklin and the Circulation of Atlantic Knowledge," in *Science and Empire in the Atlantic World*, ed. James Delbourgo and Nicholas Dew (New York: Routledge, 2008), 73–96.
10. Edmund Russell, "Evolutionary History: Prospectus for a New Field," *Environmental History* 8 (2003): 204–228.
11. On the construction of knowledge about climate change at the global scale, see also David Demeritt, "The Construction of Global Warming and the Politics of Science," *Annals of the Association of American Geographers* 91 (2001): 307–337.
12. Paul N. Edwards, "Meteorology as Infrastructural Globalism," *Osiris* 21 (2006): 229–250. See also Edwards, *A Vast Machine: Computer Models, Climate Data, and the Politics of Global Warming* (Cambridge: MIT Press, 2010); and James R. Fleming, Vladimir Jankovic, and Deborah R. Coen, eds., *Intimate University: Local and Global Themes in the History of Weather and Climate* (Sagamore Beach, Mass.: Science History Publications, 2006).
13. For two studies of biodiversity science at the national level, see Mark V. Barrow, *A Passion for Birds: American Ornithology After Audubon* (Princeton: Princeton University Press, 1998); and Michael L. Lewis, *Inventing Global Ecology: Tracking the Biodiversity Ideal in India, 1947–1997* (Athens: Ohio University Press, 2004).
14. On the creation of vegetation regions through mapping, see Malcolm Nicolson, "Alexander von Humboldt, Humboldtian Science, and the Origins of the Study of Vegetation," *History of Science* 25 (1987): 167–194. On mapping of human racial types in the field, see Jeremy Vetter, "Wallace's *Other* Line: Human Biogeography and Field Practice in the Eastern Colonial Tropics," *Journal of the History of Biology* 39 (2006): 89–123. On topographical mapping, see Matthew H. Edney, *Mapping an Empire: The Geographical Construction of British India, 1765–1843* (Chicago: University of Chicago Press, 1997).
15. For other cases of extending verticality downward, see also Helen M. Rozwadowski, *Fathoming the Ocean: The Discovery and Exploration of the Deep Sea* (Cambridge: Harvard University Press, 2005); Ronald E. Doel, Tanya J. Levin, and Mason K. Marker, "Extending Modern Cartography to the Ocean Depths: Military Patronage, Cold War Priorities, and the Heezen-Tharp Mapping Project, 1952–1959," *Journal of Historical Geography* 32 (2006): 605–626; and Naomi Oreskes, "A Context of Motivation: U.S. Navy Oceanographic Research and the Discovery of Sea-Floor Hydrothermal Vents," *Social Studies of Science* 33 (2003): 697–742.

16. Bruce Braun, "Producing Vertical Territory: Geology and Governmentality in Late Victorian Canada," *Ecumene* 7 (2000): 7–46. For another valuable study of governmentality in field science, see Scott Kirsch, "John Wesley Powell and the Mapping of the Colorado Plateau, 1869–1879: Survey Science, Geographical Solutions, and the Economy of Environmental Values," *Annals of the Association of American Geographers* 92 (2002): 548–572.
17. The role of nineteenth-century geologists in the mining industry, for example, is ably analyzed in Paul Lucier, "Commercial Interests and Scientific Disinterestedness: Consulting Geologists in Antebellum America," *Isis* 86 (1995): 245–267. The shaping of antebellum geological fieldwork by an internal improvements agenda is demonstrated in Benjamin R. Cohen, "Surveying Nature: Environmental Dimensions of Virginia's First Scientific Survey, 1835–1842," *Environmental History* 11 (2006): 37–69.
18. The role of ecologists in attempts to restore a working forest landscape is insightfully examined in Emily K. Brock, "The Challenge of Reforestation: Ecological Experiments in the Douglas Fir Forest, 1920–1940," *Environmental History* 9 (2004): 57–79. For another valuable case study revealing debates over habitat destruction, intensive human management, and the protection of biodiversity, see Peter S. Alagona, "Biography of a 'Feathered Pig': The California Condor Conservation Controversy," *Journal of the History of Biology* 37 (2004): 557–583.
19. Christine Keiner, *The Oyster Question: Scientists, Watermen, and the Maryland Chesapeake Bay Since 1880* (Athens: University of Georgia Press, 2009).
20. Cindy S. Aron, *Working at Play: A History of Vacations in the United States* (New York: Oxford University Press, 1999).
21. For example, see James A. Secord, "King of Siluria: Roderick Murchison and the Imperial Theme in Nineteenth-Century British Geology," *Victorian Studies* 25 (1982): 413–442; Robert A. Stafford, "Annexing the Landscapes of the Past: British Imperial Geology in the Nineteenth Century," in *Imperialism and the Natural World*, ed. John M. MacKenzie (Manchester: Manchester University Press, 1990), 67–89; and Suzanne Zeller, "The Colonial World as Geological Metaphor: Strata(gems) of Empire in Victorian Canada," *Osiris* 15 (2000): 85–107.
22. For a classic historical overview, see Donald Worster, *Nature's Economy: A History of Ecological Ideas* (Cambridge: Cambridge University Press, 1977).
23. Even the discourse of environmental sustainability has become the "green neoliberalism" of global financial institutions, as analyzed in Michael Goldman, *Imperial Nature: The World Bank and Struggles for Social Justice in the Age of Globalization* (New Haven: Yale University Press, 2005). For two forceful critiques of the sustainability of capitalism, see James O'Connor, "Is Sustainable Capitalism Possible?" in *Natural Causes: Essays in Ecological Marxism* (New York: Guilford Press, 1998), 234–253; and Immanuel Wallerstein, "Ecology and Capitalist Costs of Production: No Exit," in *Ecology and the World-System*, ed. Walter L. Goldfrank, David Goodman, and Andrew Szasz (Westport, Conn.: Greenwood Press, 1999), 3–11.
24. James C. Scott, *Seeing Like a State: How Certain Schemes to Improve the Human Condition Have Failed* (New Haven: Yale University Press, 1998); and Paul R. Josephson, *Industrialized Nature: Brute Force Technology and the Transformation of the Natural World* (Washington, D.C.: Island Press, 2002). For a study of the scientific guidance of development that stresses the differences between the American and Soviet examples, see Douglas R. Weiner, "The Changing Face of Soviet

Conservation," in *The Ends of the Earth*, ed. Donald Worster (Cambridge: Cambridge University Press, 1988), 252–273.

25. For a sampling focused on the field sciences, see Brian Wynne, "May the Sheep Safely Graze? A Reflexive View of the Expert-Lay Knowledge Divide," in *Risk, Environment, and Modernity: Towards a New Ecology*, ed. Scott Lash, Bronislaw Szerszynski, and Brian Wynne (London: Sage Publications, 1996), 44–83; Florian Charvolin, "Une science citoyenne? Le programme Feederwatch et la politique des grands nombres," *Développement Durable et Territoires* 19 (2004), http://developpementdurable.revues.org/index687.html; and Rebecca Ellis and Claire Waterton, "Environmental Citizenship in the Making: The Participation of Volunteer Naturalists in UK Biological Recording and Biodiversity Policy," *Science and Public Policy* 31 (2004): 95–105. For broader overviews, see Frank Fischer, *Citizens, Experts, and the Environment: The Politics of Local Knowledge* (Durham: Duke University Press, 2000); and Stephen Bocking, *Nature's Experts: Science, Politics, and the Environment* (New Brunswick: Rutgers University Press, 2004).
26. For two examples, see Alex Soojung-Kim, "Gender, Culture, and Astrophysical Fieldwork: Elizabeth Campbell and the Lick Observatory–Crocker Eclipse Expeditions," *Osiris* 11 (1996): 15–43; and Lyn Schumaker, *Africanizing Anthropology: Fieldwork, Networks, and the Making of Cultural Knowledge in Central Africa* (Durham: Duke University Press, 2001).
27. Helen M. Rozwadowski, "Small World: Forging a Scientific Maritime Culture for Oceanography," *Isis* 87 (1996): 409–429.
28. For an example from vertebrate paleontology defending this point directly, see Jeremy Vetter, "Cowboys, Scientists, and Fossils: The Field Site and Local Collaboration in the American West," *Isis* 99 (2008): 273–303.
29. Stuart McCook, "'It May Be Truth, but It Is Not Evidence': Paul du Chaillu and the Legitimation of Evidence in the Field Sciences," *Osiris* 11 (1996): 177–197.
30. Robert E. Kohler, "From Farm and Family to Career Naturalist: The Apprenticeship of Vernon Bailey," *Isis* 99 (2008): 28–56. More broadly, in thinking historically about training students in field practices, consider the late nineteenth-century flourishing of "nature study," discussed, for example, in Sally Gregory Kohlstedt, "Nature, Not Books: Scientists and the Origins of the Nature-Study Movement in the 1890s," *Isis* 96 (2005): 324–352; and Kevin Armitage, *The Nature Study Movement: The Forgotten Popularizer of America's Conservation Ethic* (Lawrence: University Press of Kansas, 2009).
31. Robert E. Kohler, *All Creatures: Naturalists, Collectors, and Biodiversity, 1850–1950* (Princeton: Princeton University Press, 2006), 184, and further developed in Kohler, "From Farm and Family to Career Naturalist."
32. Vetter, "Cowboys, Scientists, and Fossils." See also Anne Secord, "Corresponding Interests: Artisans and Gentlemen in Nineteenth-Century Natural History," *British Journal for the History of Science* 27 (1994): 383–408.
33. For an insightful critique, see Arun Agrawal, "Dismantling the Divide Between Indigenous and Scientific Knowledge," *Development and Change* 26 (1995): 413–439.
34. For a set of essays on the role of place-based learning in environmental education and life, see David W. Orr, *Earth in Mind: On Education, Environment, and the Human Prospect* (Washington, D.C.: Island Press, 1994).

One

Michael S. Reidy

From the Oceans to the Mountains

Spatial Science in an Age of Empire

Empires operate through the establishment of order. Ordering the natural environment enables imperial regimes to project power more efficiently across space, but in the process, they also quite explicitly reorganize space to their own liking, a project that helps lubricate the mechanisms of control. For this reason, Western imperial powers attempt to standardize quantities of all types, both physical and imaginary. Part of the process of Romanization, for instance, entailed the re-spatiation of newly conquered territories. Through archaeological air photography, one can still detect the Roman centuriation patterns dotting the landscape in North Africa, France, and Britain.[1] Likewise, in France after the fall of the monarchy, the new republic attempted to standardize measurements "for all people, for all time" through the mandate of the meter.[2] Napoleon, of course, not only led his armies across Europe to spread his code of law, but he also brought with him the new metric system. Centuriation and metrification were powerful transformative instruments of empire.

Britain, as the premier Western imperial power in the nineteenth century, likewise participated in the reordering of nature through the process of mapping and the establishment of trigonometrical surveys. As Matthew Edney has argued, the British hoped to "reduce India to a rigidly coherent, geometrically accurate, and uniformly precise imperial space."[3] Cartography both created an image of India as a unified and controllable territory and reinforced British notions of their own superiority. Similarly, "by ordering chaotic spaces," D. Graham Burnett wrote of the British in the South American colony of Guyana, "maps created imperial places; by making distant places visible they satisfied the scopic and gnostic drives of a conquering people; by abetting territorial control in practical ways they made colonies into large-scale Benthamite panopticons."[4] In both India and South America, the reconfiguration of space, and the imperial gaze that such a reconfiguration helped solidify, bolstered British territorial claims and racial value systems.

The practice of science helped transform previously unmapped spaces into imperial places. Britain certainly used advances in science and technology in its quest for domination and control.[5] In turn, the process of imperialism, which includes the ordering of nature through the reorganization of space and time, advanced scientific knowledge. The question remains, however, as to how exactly the imperial process transformed the sciences involved.[6] Although recent studies have resituated surveying and cartography at the center of European expansion, they have been largely concerned with particular geographies, such as India or Guyana. Less emphasis has been placed on "knowing global environments." Yet, while Imperial Britain was undoubtedly concerned with implementing its version of India on India, this was a mere corollary of the larger project of remaking entirely the horizontal and vertical dimensions of earthly space writ large. That is, empires are not built on the unique; they thrive on the universal. Order, like the enabling science, must travel to be useful.

As the essays in this volume suggest, the relatively new infusion of historical geography and environmental history into science studies has given historians of science a new set of analytical tools with which to engage questions of science and empire.[7] As the historical geographer David Harvey has noted: "Setting boundaries with respect to space, time, scale, and environment then becomes a major strategic consideration in the development of concepts, abstractions, and theories. It is usually the case that any substantial change in these boundaries will radically change the nature of the concepts, abstractions, and theories."[8] As the spatial scope of the research changes, so too does the resulting science. With this powerful framework in mind, I will show how the Victorians reorganized global environments both horizontally and vertically, and in turn, how this massive rewriting of three-dimensional space transformed the results of science. In particular, I will focus on two case studies, one horizontally covering the oceanscape, the other vertically covering the mountainscape, to demonstrate how the imperial process of viewing global environments helped shape natural philosophers' "concepts, abstractions, and theories."

Ocean Science and Horizontality

The English astronomer Edmond Halley was the first natural philosopher to combine a large-scale spatial science with the graphical representation of data. He created his first visual graphs to represent geophysical forces over the world's oceans. After sailing on two scientific voyages on the Atlantic between November 1698 and June 1700, he broke from the tradition of using numerical tables and chose to represent his observations of the earth's magnetic field visually as a chart. He followed his visual representations of terrestrial magnetism with a similar synoptic chart of the tides as they progressed in the Atlantic Ocean. Not content to keep his researches confined to the surface of the sea, he also studied the properties of the atmosphere above the world's oceans, publishing a "scheme, shewing at one view all the various Tracts and Courses of these Winds."[9] As one of the first philosophers to study the large-scale geophysical properties of the oceans, Halley stressed the power of his new method of data representation. His synoptic charts, he argued, by enabling "one View to represent the whole,"[10] were far more useful than "any verbal description whatsoever."[11] His integration of the vertical and horizontal realms, moreover, had profound consequences on the way future natural philosophers studied the sea-air interface.

Halley's advances in both the geophysical sciences and their graphical representation lay dormant for over a century. His explicitly scientific voyages, though revolutionary, were isolated events. Not until the early nineteenth century did philosophers begin to coordinate large-scale geophysical data-gathering initiatives. As observational data from remote areas became more accessible owing to European overseas expansion, natural philosophers adopted Halley's graphical charts as the most efficient means of organizing massive amounts of data distributed over large geographical areas. The Prussian mining engineer and explorer Alexander von Humboldt, for instance, advanced on Halley's graphical techniques and extended their scope to many other areas in physical astronomy and natural history.[12]

Humboldt strove to find mathematical laws that related diverse but interconnected natural phenomena over spatially distributed areas. Only by traveling inland from the coast and collecting observations from around the globe, he argued, would natural philosophers be able to find order in nature's complexity.[13] Humboldt's spatial science heavily influenced traveling naturalists such as Charles Darwin, who developed his theory of evolution through natural selection at least in part by comparing geographically diverse observations within continents, between continents, and among different islands in the Galapagos.[14] Owing to the success of Humboldt's spatial approach in natural history, philosophers hoped similar methods would also advance the study of the physical properties of the sea. The geographies of collection that natural philosophers used to study the ocean mimicked those used by naturalists. Thus, while Darwin trekked through the inland regions of South America,

Robert Fitzroy, the scientifically minded captain of the *Beagle*, looked outward over the large expanse of the world's oceans in an attempt to order its tides and currents. As the number of vessels traveling the oceans increased, ocean scientists adopted both a geographical approach to data collection and the graphical method of representation. The Magnetic Crusade, global meteorology, and the nineteenth-century quest to understand the world's tidal oscillations and ocean currents can all be viewed as attempts to apply Humboldt's spatial science to bring the world's oceans to rule.

The growing importance of the ocean to burgeoning imperial nations compelled scientists to take a renewed interest in the physical properties of the sea. As one writer noted in the *Illustrated London News*, the ocean served the British traveler just as the atmosphere served the wings of a bird: "On its yielding bosom he now sails or steams quickly, and with ease, withersoever he will. The great highway of nations, as the sea has in consequence been appropriately called, has its own laws, which he must study to use it advantageously for his purposes. But . . . only lately have these great phenomena been considered worthy of scientific observation."[15] The writer went on to observe that mariners had previously "no more thought of studying its phenomena than mail-coachmen thought of studying astronomy or natural history as they passed on the roads." That changed significantly in the first half of the nineteenth century, a transformation of scientific perspective that conformed to the larger geopolitical ambitions of maritime nations.

A renewed interest in the study of the tides in the early nineteenth century exemplifies this heightened attention to the study of the sea.[16] Though Isaac Newton successfully used his law of universal gravitation to explain how the forces of the sun and moon produced the tides in the ocean, he failed to translate his theory into concrete methods of tidal prediction. That is, though Newton's tidal theory was correct in principle, it was far too general to work in practice.[17] In a prize competition adjudicated by the French Academy in 1740, the renowned mathematician Daniel Bernoulli advanced significantly on Newton's theory, offering rules to construct tide tables at specific ports once enough observations had been gathered. In the early nineteenth century, this was known as the "Newton-Bernoulli equilibrium theory" of the tides. To use the equilibrium theory to produce accurate tide tables, however, researchers required long-term observations of the tides at each and every port. Thus, when John William Lubbock, a London banker and Cambridge graduate, took up the study of the tides in 1829, he attempted to find long-term observations of the tides at all the major ports in Britain. He managed to acquire twenty-five years of observations from the London Docks Company and nineteen years from the Athenaeum in Liverpool. He used these observations to calculate the major tidal constants needed for the equilibrium theory, which in turn allowed him to produce relatively accurate tide tables, but only for those two ports. He had a difficult time acquiring other tidal data. Natural philosophers rarely made observations themselves; rather, dockyard officials, tide table makers, and harbor masters collected tidal readings, which they justifiably viewed as their own private property. They used the observations to produce their own private tide tables, which they then marketed for profit in the bustling port cities throughout Great Britain.

Though Lubbock failed to procure much in the way of long-term data, he did succeed in convincing his former tutor, William Whewell, to take up the science of the tides. Whewell graduated from Trinity College, Cambridge, as second Wrangler in 1826, allowing him to stay at Cambridge throughout his career. He eventually accepted the Mastership of Trinity in 1841, a Crown appointment he held until his death in 1866. When he became involved in tidal studies (which he termed "tidology"), he held a professorship in mineralogy at Cambridge, intent on making a name for himself as a practicing scientist (a term he also coined).[18] As early as 1833, he set out a long-term research program in tidology that he would follow, intermittently, for over two decades.

In the last section of his first publication on the tides, the paper sent to Fitzroy while he was on board the *Beagle*, Whewell outlined two distinct approaches to the subject.[19] The first described Lubbock's work to find the major tidal constants for specific ports through long-term observations. "But in the meantime," Whewell argued, "no one appears to have attempted to trace the nature of the connexion among the tides of different parts of the world."[20] Rather than viewing the tides temporally and relying on long-term observations, as Lubbock had done, Whewell's second approach entailed studying them spatially over large geographical areas, relying on short-term but connected observations acquired from as many ports as possible. As Whewell put it: "Continued observations at the same place are connected by relations of time; comparative observations at different places are connected by relations of space. The former relations have been made the subject of theory, however imperfectly: the latter have not."[21] In essence, he intended to transform the study of the tides into a geographic, spatial science.

Such an approach had its advantages. First, Whewell did not have to rely on the acquisition of long-term observations from commercially inclined tide table makers. Second, and most important for Whewell, it solved crucial problems that had arisen in the equilibrium theory. The forces of the sun and moon in the open ocean produce the tides that hit the coasts and estuaries of Europe. A lag of time exists, however, between the action of the tidal generating forces as they act in the open ocean and their eventual flux and reflux on the coasts of

Europe. Lubbock and Whewell referred to this lag of time as the "age of the tide." To determine the forces that produced the tides in each port, therefore, equilibrium theorists used the position of the sun and moon at some previous time or "epoch." But the equilibrium theory offered no help in the choice of epochs. Indeed, Lubbock spent most of his time in tidology trying to figure out which epoch to use for each tide generating force. He had professional calculators begin with the epoch contemporaneous with the tide as it rolled onto the coasts of England, and then had them work backward in increments of twelve hours. His calculators produced tables for each twelve-hour period, which they subsequently compared directly with observations. Lubbock then used the epoch that most closely fit the observations, an empirically based method of saving the phenomena that had no theoretical justification.

For Whewell, whose historical and philosophical works focused heavily on the fundamental role of theory in directing the scientific process, this lack of theoretical guidance in the study of the tides led him to his new spatial approach. He realized that comparative observations, if gathered and exhibited correctly, could solve the problem of finding the "epochs" used in the equilibrium theory. Short-term, comparative observations, Whewell noted, were almost impossible to reduce to calculation owing to the "extreme complexity" of forces on which they depended. "But, though the connexion of the tides in different places cannot be calculated," Whewell argued, "it can be expressed." While long-term observations could be put into tables, comparative observations required the use of the graphical method. Whewell created what he termed "cotidal" maps, similar to the charts first introduced by Halley in the seventeenth century and advanced by Humboldt in Whewell's own lifetime. If the correct epochs for only one port were known with certainty, Whewell reasoned, by tracing lines through all parts of the coast and open ocean that experienced high water at the same time, he could determine the epochs of all other ports simply by counting the intervening lines. "For the age of the original tide in any part of the open ocean being known," Whewell explained, "the age of the tide derived from the original tide in any other part would be known from the number of intervening cotidal lines."[22] Like Halley a century and a half earlier, Whewell highlighted the advances such a graphical method produced. His spatial approach combined with a new way of expressing the data would enable him "to draw a map of cotidal lines with certainty and accuracy; and thus to give, upon a single sheet, a tide table for all ports of the earth."[23] Just as naturalists created maps of the territory in which they collected, and geologists graphically represented the strata of the earth, so too could Whewell create a synoptic view of the world's tides on a single sheet of paper.

For Whewell's comparative approach to work, however, the actual gathering of short-term observations from around the globe had to be simultaneous. Whewell realized that he would require the financial support and imperial connections of the British government. From the very beginning of his tidal studies, Whewell worked closely with Francis Beaufort, the hydrographer to the Admiralty. Like Fitzroy and other scientific servicemen, Beaufort had kept up with advances in the sciences as a way of furthering his own career. Indeed, as the king's hydrographer, Beaufort was in charge of keeping the ocean safe through the production of charts, a process that relied heavily on advances in surveying and cartography. With Whewell's help, Beaufort arranged to have officers of the Preventive Coast Guard stationed at over five hundred places around the coasts of Great Britain collect tide observations simultaneously for a fortnight in June 1834. After the successful completion of what Whewell termed his "great tide experiment," Beaufort suggested that a similar experiment should be extended beyond the British Isles to encompass the entire globe.

In June of the following year, hydrographers or their counterparts in nine countries—the United States, France, Spain, Portugal, Belgium, Denmark, Norway, the Netherlands, and Great Britain and Ireland—ordered their subordinates to measure the tides simultaneously on the coasts of their countries and possessions. Observers, some equipped with sophisticated tide gauges, took measurements of the tides every fifteen minutes at over 660 tide stations around the Atlantic and, to a lesser extent, the Pacific Oceans. Beaufort also contacted all the surveyors under his command to have the tides observed for the entire month as part of their official duties. Data arrived from Table Bay and Simon's Bay in South Africa, from the principal ports in Australia and New Zealand, and from numerous islands where British officers were stationed, including the Isle of Man, Mauritius, Malta, Ceylon, and several of the Channel Islands. Whewell then produced a cotidal map (see figure 1.1) that visually represented how the tides progressed throughout the world's oceans. It was printed, as Whewell had foreseen, on a single sheet of paper.

While the cotidal map perfectly served Whewell's spatial approach to science, the graphical method also nicely complemented the Admiralty's own designs. It offered, in one synoptic glance, the exact information the Admiralty needed to get their vessels safely in and out of distant and often dangerous ports. Moreover, that Beaufort took charge of the multinational investigation should come as no surprise. The Admiralty profited most from Whewell's spatial approach to the study of the tides, an approach that would have been impossible without Britain's imperial reach.

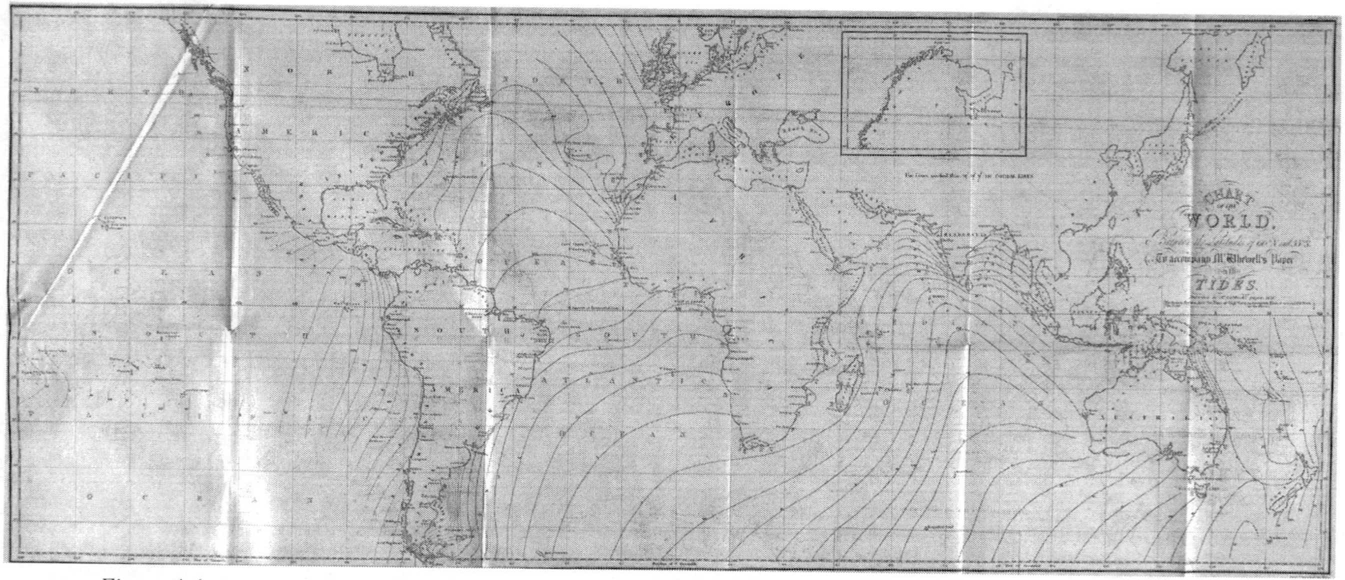

Figure 1.1

A chart of the cotidal lines of the world's oceans.

From William Whewell, "Essay Towards a First Approximation to a Map of Cotidal Lines," Philosophical Transactions of the Royal Society of London *123 (1833): 147–236.*

Mountain Science and Verticality

As the British Empire expanded to distant regions of the globe, so too did its enabling science. Indeed, as Whewell's tidology demonstrates, Britain's expanding empire enabled Whewell to change the methods and scope of his tidal analysis. Researchers in other areas of geophysics, such as terrestrial magnetism and meteorology, also quite consciously rode the imperial wave to advance their science through the acquisition of data from around the world. Following Humboldt, these researchers created iso-maps of all types in an attempt to bring order to the watery frontier.[24] Yet, similar to Halley a century earlier, Humboldt also investigated the properties of the atmosphere, which he believed interacted with the tides and currents of the ocean and the flora and fauna on the land. Combined with his lifelong fascination with volcanoes and the distribution of mountain ranges, his theory of the interrelation among diverse forces in nature proffered a vertical orientation to many of his musings.[25] His visual representations of the earth's vegetation zones, for instance, had a profound influence on nineteenth-century British natural philosophers. Naturalists took this verticality into account in formulating their research agendas.

Joseph Dalton Hooker was one such naturalist. He became interested in natural history through his father, William Jackson Hooker, who had traveled and collected throughout England and Scotland, in Europe, and eventually in Iceland as part of a state-sponsored voyage of scientific discovery. Appointed in 1841 to direct the Royal Botanic Gardens at Kew, William transformed the institution into a center of economic botany. Joseph Hooker received from his father a penchant for both travel and botany. After attaining his medical degree from the University of Glasgow at the age of twenty-one, Joseph prepared himself for a career as a botanist. By the early nineteenth century—as evidenced by the experiences of Humboldt in South America, Darwin on the *Beagle*, the naturalist T. H. Huxley on the *Rattlesnake*, and the biogeographer Alfred Russel Wallace in the Amazon—postgraduate education frequently took the form of travel to distant areas, often as part of governmental surveys. "From my earliest childhood," Hooker later confided to Darwin, "I nourished and cherished the desire to make a creditable Journey in a new country . . . as should give me a niche amongst the scientific explorers of the globe I inhabit, and hand my name down as a useful contributor of original matter."[26] Hooker participated in two major voyages of scientific discovery in his attempt to establish his career as a botanist.[27]

On his first voyage, Hooker served as assistant surgeon and naturalist accompanying the celebrated explorer Sir James Clark Ross to the Antarctic between 1839 and 1842. The young Hooker had met Charles Darwin briefly before his departure, and the two aspiring naturalists began to correspond regularly after his return. Indeed, Darwin intimated his thoughts on evolution to Hooker after Hooker was settled comfortably back in London, and right away Hooker began contemplating a second sojourn, this time specifically to the mountains. Science followed the Crown, and the Crown had particular interests in India, so, on the suggestion of Lord Auckland, 1st Lord of the Admiralty, and Hugh Falconer, director of the Calcutta Botanic Garden, Hooker began to plan a second voyage of discovery to collect botanical treasures in the Himalaya.

The correspondence between Darwin and Hooker covers a wide range of topics in both botany and geology, but by far the most prevalent is the question of the geography of plants, particularly the extent of species and barriers to their migration.[28] Both naturalists, it appears, were interested in questions of altitude in relation to plant distribution owing to the seminal work published several decades earlier by the indefatigable Humboldt, whose "Essay on the Geography of Plants" (1805) established the scientific framework for studying the relationship among climate, elevation, and the distribution of species.[29] Humboldt had begun writing the paper during his ascent to the perpetual snows of Mount Chimborazo, the highest peak in Ecuador and then thought to be the highest peak in the world (see figure 1.2). The aim of his essay was to convince other naturalists that the new science of plant geography was an "essential part of general physics," and within its pages he introduced the major questions that European and American naturalists would investigate for the next century.[30]

Studying the geography of plants, Humboldt claimed, could help scientists recognize when islands had separated from each other and from major continents, and whether this occurred before or after the development of organisms. It could also help naturalists comprehend "whether . . . the whole surface of the world was immediately covered by a diversity of plants, or if . . . the globe at rest only produced plants in one region which sea currents then transported over the centuries to progressively more remote zones."[31] These questions of migration and barriers to migration, and of single or multiple centers of creation, were the exact questions that Hooker and Darwin pursued throughout their careers. Moreover, in a passage that still forms one of the main questions of modern ecology, Humboldt linked the study of biogeography directly to human agency: to wars, passions, and the human diaspora throughout the globe. Whereas winds, currents, and birds could "aid the migration of plants," Humboldt argued, it was "man [that] primarily takes care of this."[32] The human predilection to cultivate newly introduced plants, for instance, always curtailed the spread of native species. "These are some of the considerations

Figure 1.2

Alexander von Humboldt's depiction of zones of vegetation on Mount Chimborazo (ca. 1805).
From Wikimedia Commons.

agriculture presents, and its various produce depends on the latitude, origin, and needs of people. The influence of food, more or less stimulating the character and energy of passions, naval history, and wars undertaken for the dispute of produce of the vegetable kingdom; these all link the Geography of Plants to the political and moral history of man."[33] Owing to the overtly imperial nature of Hooker's own studies in the Himalaya, Humboldt's words proved particularly prophetic (though they seemed to have been lost on Hooker himself).

In the years preceding Hooker's Himalayan tour, Darwin and Hooker argued over Humboldt's seemingly straightforward relation between vegetation and altitude. Hooker thought the distribution of plants on mountains had more to do with moisture in the air, atmospheric pressure, and perhaps with the radical changes in temperature that high-altitude environments experience. "A mountain immediately gives new vegetable forms," he wrote to Darwin, "perhaps not so much because its temperature is lower but because its vicissitudes are more remarkable."[34] Hooker was interested in the connection between climate and species for purposes of economic botany, but he was also intent on helping Darwin work through his secret hypothesis: "Nothing will give me so much pleasure as to get grounds for your reasonings & carry out your theory of isolation."[35] In the meantime, owing to Darwin's repeated insistence, Hooker was almost convinced to "consider migration as the only cause of the dispersion or diffusion of so called species."[36] Yet he needed more data, and for that he needed to travel to the mountainous regions himself: "Should I be able to trace the majority of them from one zone to the other, I shall declare myself a good migrationist, if not I must hold the question still unsettled."[37]

Subsequent correspondence between Darwin and Hooker began to focus more particularly on the dispersion of mountain flora. The timing was propitious because Hooker had succeeded in procuring funding from the government for his travels to India, home to the highest mountains in the world. European naturalists had failed to study the flora and fauna of the Himalaya because of their extreme climate and supposed unfriendly natives. "We were ignorant even of the geography of the central and eastern portions of these mountains," Hooker lamented, "while all to the north was involved in a mystery equally attractive to the traveler and the naturalist."[38] The sheer vertical expanse of the mountains and their untouched character were what excited Hooker most about his pending trip to India. Indeed, Hooker was drawn to the extremes. After his successful voyage of discovery to Antarctica, Hooker went to the Himalaya because the region was decidedly different, because European naturalists had yet to collect in Sikkim and Tibet, and because there one could climb the highest mountains in the world. Though he changed the wording in his published journals, in his correspondence Hooker makes clear that his main goal was to collect as high as possible in "the snows."[39]

Hooker's vertical mentality and his desire to gather material for Darwin are plainly evident in his highly successful *Himalayan Journals*, a publication he dedicated to Darwin. The Himalaya enabled Hooker to explore and compare three distinct vegetation zones: a tropical zone to 6,500 feet, a temperate zone to 11,500 feet, and an alpine zone reaching to the perpetual snows.[40] This allowed him to focus on the relationship between altitude and plant life in a narrative that followed traditional travel and adventure narratives of the time. He described the landscapes, the indigenous cultures, and the material products of the lands he visited, including an in-depth analysis of the process of poppy cultivation. The narrative itself, however, is divided into three challenging mountaineering treks—two to the Himalaya and one to the Khasia Mountains—in geographical areas previously unmapped by Europeans.

He spent the summer months (the rainy season) of his first year planning his initial trek into the mountains. He stayed near Darjeeling as a guest of Brian Hodgson, an avid mountaineer who trained Hooker in the art of trekking. "The view from his windows," Hooker related, "is quite unparalleled for the scenery it embraces, commanding confessedly the grandest known landscape of snowy mountains in the Himalaya, and hence in the world. Kinchinjunga (forty five miles distant) is the prominent object, rising 21,000 above the level of the observer out of a sea of intervening wooded hills."[41] Kinchinjunga, which Hooker thought was the highest mountain in the world, is actually the third highest mountain, with five peaks, the tallest of which is 28,170 feet. To get there, Hooker proposed to travel through the Nepalese passes near Tibet, which would bring him "as near to the central mass and loftiest parts of the eastern flank of Kinchinjunga as possible."[42] With travel plans finalized and over fifty persons marshaled for the trek, the trip to the foothills of Kinchinjunga was finally underway.[43]

As the loftiest mountains were perpetually shrouded in clouds and mist, he described with ecstasy the few moments when he was able to catch a glimpse of their majestic beauty. Yet he always mixed the sublimity of the view with the equally alluring change in the landscape: "Every feature, botanical, geological, and zoological, is new on entering this district. The change is sudden and immediate; sea and shore are hardly more conspicuously different."[44] At every juncture, Hooker focused particularly on the fascinating changes in the vegetation as he traveled from valleys to glaciers: "From the deep valleys choked with tropical luxuriance to the scanty yak pasturage on the heights above, [the vegetation] resolves itself into five belts: 1, palm and plantain; 2, oak and laurel; 3, fir; 4, rhododendron and grass; 5, rock and snow. From the bed of the Ratong, in which grow palms and screw pine and plantain, it is only seven miles in a direct line to the perpetual ice. . . . All the intermediate phases of vegetation are seen at a glance."[45] Like Whewell's synoptic representations of the ocean, Hooker was attempting to view all the zones of vegetation "at a glance."

The first volume of Hooker's narrative climaxes with this first aerial ascent into the Himalaya. Volume 2, likewise, begins with his return to Darjeeling and his preparation for yet another expedition "to the loftier parts of Sikkim."[46] And though Hooker may have simply wanted to go higher than anyone else, this was not the rationale he gave in his *Journal* or his correspondence. Rather, he described his aim as purely botanical. "I have a set of most curious new plants from between 17 and 19,000 feet," he wrote back to Britain. "They are extremely scarce and require close hunting. Sometimes I get but one or two specimens of a kind, and poking with a headache is very disagreeable."[47] The extreme limits of vegetation in the Himalaya had never been studied, and Hooker was continually excited about finding new species, irrespective of his own altitude sickness.

Nowhere is his focused search for new plants at high altitudes clearer than in his second ascent into the Himalayan mountains. Hooker climbed the Donkia Pass, where he found several species of lichen: "Donkia is a wonderful place; 19,200 feet is the altitude of the Pass, and plants to 200 feet of top, Lichens to all but 20,000 feet."[48] He was filled with excitement, as he had seen the same lichen while cruising the coast of Antarctica, on Cockburn Island, as a member of Ross's crew: "I was greatly pleased with finding my most Antarctic plant, *Lecanora miniata*, at the top of the Pass, and today I saw stony hills at 19,000 feet stained wholly orange red with it, exactly as the rocks of Cockburn Island were in 64 degrees South; is not this most curious and interesting? To find the identical plant forming the only vegetation at the two extreme limits of vegetable life is always interesting."[49] It was both curious and interesting because he believed it would shed light on the laws of the distribution of species.

After his successful attempt to climb well above nineteen thousand feet—perhaps higher than any other European had ever climbed—the narrative changes to one of imperial adventure, a topic that has excited past historians.[50] From the Donkia Pass, Hooker made an illegal and foolhardy jaunt into Tibet, outriding the Sikkim guards sent to the border to deter him. His violation of entering Tibet placed the Sikkim Rajah in a difficult position, as he was perpetually fearful of angering his Chinese neighbors. Upon Hooker's return, the Sikkim authorities arrested Hooker's climbing companion, Archibald Campbell, who alerted Hooker by yelling, "Hooker! Hooker! the savages are murdering me!"[51] Campbell was bound, beaten, and tortured. Though Hooker was never actually arrested, his guides were bound and placed in stocks, and Hooker was "retained." The result, moreover, was typical of British imperial rule. Campbell and Hooker were eventually freed, none too worse for the wear, and the British used the episode to annex further territory from Sikkim.

This second climb and subsequent arrest form the climax of the second volume of his *Himalayan Journal*. He had collected over three thousand species of plants while mountaineering in Sikkim and its environs, and many more in his subsequent travels to the Khasia Mountains.[52] Hooker's treks into "the snows" enabled him to explore more fully the complicated science of biogeography. By studying the zones of vegetation, he amassed enough data to help Charles Darwin work through his own theories of plant migration. In long letters addressed to both his father and to Darwin, Hooker always emphasized the slight changes that occurred in plants as he traveled up the sides of mountains. He noted, for instance, how distinct species, though they changed slightly with elevation, also retained their similarities to the same species found on the tops of different mountain ranges, whether in the Sikkim Himalaya or in other mountain ranges around the world.

The geologist Charles Lyell and naturalist Edward Forbes had published remarkably clear and simple solutions to the problems of species migration based on dramatic climactic changes and equally remarkable changes in the levels of land and sea. To expand on their theory and link it to his (actually Hooker's) data, Darwin offered in his *On the Origin of Species* an imaginary condition in which a glacial period slowly arrived and slowly passed away. As the cold arrived, Darwin explained, "alpine inhabitants would descend to the plains," and the lowlands would be uniformly filled with arctic species. As the warmth returned and the snow and ice melted, gradually receding up the mountain, "the arctic forms would seize on the cleared and thawed ground, always ascending higher and higher." What would be left would be arctic species found only on the summits of mountains. "On the Himalaya," notes Darwin in *Origin*, his debt to Hooker unveiled, "and on the isolated mountain ranges of the peninsula of India, on the heights of Ceylon, and on the volcanic cones of Java, many plants occur, either identically the same or representing each other, and at the same time representing plants of Europe, not found in the intervening hot lowlands."[53] In the same manner, the arctic flora would also recede from the northern regions of the globe as the glacial period arrived, and return once the glacial period subsided, thus explaining the presence of both arctic plants in the higher latitudes and their similarity to plants on mountain peaks throughout the globe. This is, of course, exactly what Hooker had observed from a combination of his voyage to the Antarctic with Captain Ross and his treks for several years in the heights of the Himalaya.

From Hooker's data amassed in the Himalaya, Darwin could confidently argue that one finds both on the mountain summits and in the polar regions exactly the types of plant variations one would expect to find according to his theory: "Hence we see that through the world, the plants growing on the more lofty mountains . . . are sometimes identically the same; but they are much more oftener specifically distinct, though related to each other in a most remarkable manner."[54] Darwin used Hooker's vertical data gathering in support for his theory of evolution through natural selection. Hooker could never have helped Darwin in this manner unless he had actually climbed to the tops of mountains, carefully collected and labeled plant specimens based on their altitude, and then compared them to plants collected on other mountain ranges, which is why Hooker hightailed it to the tallest mountains whenever and wherever he traveled.

Conclusion

In the final few pages of *Civilization and Its Discontents*, Sigmund Freud suggested that the development of civilizations shared a "far-reaching similarity to the development of the individual."[55] He supported this analogy by noting the comparable means by which individuals and civilizations attempt to delimit the "potent obstacle" of aggressiveness. The establishment of order figured prominently in the process. "Order," Freud opined, "is a kind of compulsion to repeat which, when a regulation has been laid down once and for all, decides when, where and how a thing shall be done so that in every similar circumstance one is spared hesitation and indecision. The benefits of order are incontestable. It enables men to use space and time to the best advantage, while conserving their psychical forces." Freud placed "man's observation of the great astronomical regularities" as both the point of departure and the model for such regulating processes, a list that he then extended to all humanity's higher mental faculties, including "intellectual, scientific, and artistic achievements." Though Freud was ultimately unhappy with this "dangerous" analogy between individuals and the state, he needed not hedge on this point. From mapping the heavens to mapping the oceans and mountains, Western imperial nations used science to establish order through the reconstruction of geographical space, a process that enabled empires to project power more efficiently without "hesitation and indecision." Throughout the nineteenth century, natural philosophers participated in a new spatial approach to investigating nature's laws, one that covered vast horizontal and vertical distances. They remade these spaces throughout the process, bounding oceans with iso-lines and mountains with vegetation zones. Maps, whether iso-tidal or biogeographical, enabled imperial nations, in Freud's words, to "use space and time to the best advantage."[56]

Whewell's great achievement in tidology was to transform the study from a temporal to a spatial science. Unfortunately for Whewell and his science, such an approach does not work for understanding the tides. Tidal action, though influenced initially by the forces of the sun and moon, is primarily a local phenomenon, based on the hydrodynamics of waves on local coasts. To determine the tides on most shores, one needs to study not only astronomical forces, but also the shape of the estuary or coast, the depth of the water, and the direction of the wind. What is most striking is how glaringly obvious this is on the coasts of Europe. Indeed, Whewell realized during his initial tide experiment around Great Britain that the same astronomical forces produced extraordinarily different tides, even nearby on the same coast. The imperial process, however, led Whewell to overlook this blatant fact. With the mighty force of the imperial British Navy as one of his instruments, Whewell deliberately forced a global framework onto the tides. Changing the spatial boundaries of the study was thus a strategic decision on the part of the British scientific and military elite, one that readily fit the resources of the British Admiralty leading to a more useful conceptualization of the oceanic environment. They wanted to be able to view the tides on a single sheet of paper. This seemingly benign desire, in turn, shaped the nature of Whewell's "concepts, abstractions, and theories." In short, Whewell made the tides conform to a worldwide framework because the British Empire excelled at the global, not the local, level. Whewell's tidology followed the flag.

Hooker's botanical research also followed the flag; or rather, the flag resolutely followed Hooker. He traveled to remote and unmapped areas in the Himalayan frontier where he could view firsthand the transition of flora from tropical to temperate to arctic zones. In the process, he was able to advance on Humboldt's plant geography in significant ways. Humboldt had related plant life to elevation based on mean temperatures, whereas Hooker found that the moisture content of the air and the extreme vicissitudes between hot and cold were the determining factors. He had reached this conclusion from having traveled both to higher latitudes and to higher altitudes, specifically focusing on the parallels between them. Hooker's imperial wonderings, funded by the Crown, determined the geographical scope of his studies, which in turn shaped the nature of his "concepts, abstractions, and theories." Moreover, by climbing not only in the Himalaya, but also in other mountain ranges around the globe, he was able to confirm for Darwin the vertical migration and distribution of plants species. And finally, his studies in geographical botany eventually convinced him that Darwin's theory of transmutation was indeed correct. By the time Darwin published, he had the top botanist in Britain on his side.

Humboldt quite correctly linked the geography of plants to the "political and moral history of man." Imperial might and the value systems it engendered helped sustain economic botany throughout the Victorian era. For Hooker, this meant that he was justified in his illicit jaunt into the heights of Tibet, a move that led to his climbing partner's arrest and torture. He traveled as an agent of the Crown, and the relationship between his collecting and British imperialism was explicit. He wrote detailed letters to his father at Kew concerning the possibilities of cultivating cotton, tobacco, sugar, and other plants, and he used his skill as an artist to produce a map of Sikkim, published by the Indian Trigonometric Survey and reproduced in his own *Himalayan Journals*. Mapping the country for the good of Britain was always on his mind. "I had just finished for you an excellent large map of my wanderings," he wrote to Darwin while in Sikkim, "but have thought it proper to give it to Gen. Young, who was all abroad as to how to dispose of the troops now marching into Sikkim."[57] Although Hooker fails to mention this to Darwin, troops were on the march because of Hooker's disgraceful disregard of the Sikkim authorities during his sojourn into Tibet.

Expansionist nations required a scientific understanding of the spatial relations of their territories. Throughout the middle decades of the nineteenth century, natural philosophers filled the oceans with iso-lines of all types, a graphical and imperial project that helped define the ocean as a safe and controllable environment. The end product was an ordered ocean fit to sail. Naturalists during this same period similarly bounded the mountainside with zones of all types, from meteorological to botanical. Humboldt, Hooker, and others had studied vegetation zones at specific sites in South America, India, and elsewhere, but when transferred to Darwin's *Origin*, these zones appeared universal and ordered. Vegetation on all mountains, everywhere in the world, followed a lawlike pattern of change, powerful evidence according to Darwin of the transmutation of species owing to the mechanism of natural selection. Vertical and horizontal spaces were created anew, a conceptualization that legitimized both the spatial turn in science and the expansionist programs of the enabling imperial power.

Sometimes the reorganization of space that accompanied empire proved extremely useful to science, as was the case with biogeography; at other times, it proved misleading, eclipsing useful conclusions for almost half a century, as was the case with the tides. It would be impossible at the time to tell which was occurring. Yet one outcome is certain: science itself profited every time. And the practice of science always supported the imperial process, regardless of whether the theories were correct. Science helped imperial nations like Britain create ordered spaces in the natural world—big spaces, wide spaces, spaces of the earth, spaces of empire—which then directed the actions of scientists, in how they created their concepts, abstractions, and theories. Returning full circle, imperial nations then took from science those concepts and ideas, no matter how real or imagined, that supported their expansionist programs, making empire appear legitimate, rational, even natural. Want to know one way to "know global environments"? Follow the British scientists in the mid-nineteenth century as they sailed the waves of the open ocean or climbed the vertical cliffs of distant mountains.

Notes

1. Centuriation is "the division of the land into *centuriae* each of 200 *iugera*, roughly 700 metres each way." Colin Wells, *The Roman Empire*, 2nd ed. (Cambridge: Harvard University Press, 1992), 50, 162.
2. Ken Alder's award-winning book is aptly titled *The Measure of All Things: The Seven-Year Odyssey and Hidden Error That Transformed the World* (New York: Free Press, 2002), 1.
3. Matthew H. Edney, *Mapping an Empire: The Geographical Construction of British India, 1765–1843* (Chicago: University of Chicago Press, 1990), 319.
4. D. Graham Burnett, *Masters of All They Surveyed: Exploration, Geography, and British El Dorado* (Chicago: University of Chicago Press, 2000), 6.
5. See, for example, Lucile H. Brockway, *Science and Colonial Expansion: The Role of the British Royal Botanical Gardens* (New York: Academic Press, 1979); John M. MacKenzie, *Imperialism and the Natural World* (Manchester: Manchester University Press, 1991); Roy MacLeod, ed., *Nature and Empire: Science and the Colonial Enterprise, Osiris* 15 (2000); Alex Soojung-Kim Pang, *Empire and the Sun: Victorian Solar Eclipse Expeditions* (Stanford: Stanford University Press, 2002); and Richard Drayton, *Nature's Government: Science, Imperial Britain, and the "Improvement" of the World* (New Haven: Yale University Press, 2000).
6. For an early but interesting debate, see Lewis Pyenson, *Cultural Imperialism and Exact Sciences: German Expansion Overseas, 1900–1930* (New York: Lang, 1985), and his *Empire of Reason: Exact Sciences in Indonesia, 1840–1940* (Leiden, Netherlands: Brill, 1989); the critique by Paolo Palladino and Michael Worboys, "Science and Imperialism," *Isis* 84 (1993): 91–103; and Pyenson's response, "Cultural Imperialism and Exact Sciences Revisited," *Isis* 84 (1993): 103–108.
7. Janet Browne, "Biogeography and Empire," in *Cultures of Natural History*, ed. N. Jardine, J. A. Secord, and E. C. Spary (Cambridge: Cambridge University Press, 1996), 305–321; Robert E. Kohler, *Landscapes and Labscapes: Exploring the Lab-Field Border in Biology* (Chicago: University of Chicago Press, 2002); Kohler, "Place and Practice in Field Biology," *History of Science* 40 (2002): 189–210; Gregg Mitman, "Hay Fever Holiday: Health, Leisure, and Place in Gilded-Age America," *Bulletin of the History of Medicine* 77 (2003), 600–635; David N. Livingstone, *Putting Science in Its Place: Geographies of Scientific Knowledge* (Chicago: University of Chicago Press, 2003). Simon Naylor, "Introduction: Historical Geographies of Science: Places, Contexts, Cartographies," *British Journal for the History of Science* 38 (2005): 1–12; and Megan Raby, "Making Science Travel: Geographies of

Collection and the American Ornithologists' Union Code of Nomenclature" (master's thesis, Montana State University, 2007).
8. David Harvey, *Justice, Nature, and the Geography of Difference* (Cambridge, Mass: Blackwell, 1996), 53.
9. Edmund Halley, "An Historical Account of the Trade Winds, and Monsoons, Observable in the Seas between and near the Tropicks, with an Attempt to Assign the Physical Cause of the Said Winds," *Philosophical Transactions* 16 (1686–1692): 153–168. For a detailed account of Halley's voyages of scientific discovery, see Alan H. Cook, *Edmond Halley: Charting the Heavens and the Seas* (New York: Clarendon Press, 1998); and Norman J. Thrower, ed., *The Three Voyages of Edmond Halley in the Paramore, 1698–1701* (London: Hakluyt Society, 1981).
10. Halley to Burchett, 23 April 1701, in *Correspondence and Papers of Edmond Halley*, ed. Eugene Fairfield MacPike (1932; repr., Oxford: Clarendon Press, 1975).
11. Halley, "An Historical Account," 163.
12. Nicolaas Rupke, "Humboldtian Distribution Maps: The Spatial Ordering of Scientific Knowledge," in *The Structure of Knowledge: Classification of Science and Learning Since the Renaissance*, ed. Tore Frängsmyr (Berkeley: Office for History of Science and Technology, University of California, 2001), 93–116.
13. Michael Dettelbach, "Humboldtian Science," in Jardine, Secord, and Spary, *Cultures of Natural History*, 287–304.
14. Susan Faye Cannon, *Science in Culture* (New York: Science History Publications, 1978). According to Cannon, Darwin was the quintessential Humboldtian.
15. Anon., *Illustrated London News*, April 14, 1855.
16. For an in-depth account of the study of the tides in Britain in the nineteenth century, see Michael S. Reidy, *Tides of History: Ocean Science and Her Majesty's Navy* (Chicago: University of Chicago Press, 2008).
17. For Newton's treatment of the tides, see Margaret Deacon, *Scientists and the Sea* (London: Academic Press, 1971); and David Edgar Cartwright, *Tides: A Scientific History* (Cambridge: Cambridge University Press, 1999).
18. For Whewell, see Richard Yeo, *Defining Science: William Whewell, Natural Knowledge, and Public Debate in Early Victorian Britain* (Cambridge: Cambridge University Press, 1993); Menachem Fisch, *William Whewell, Philosopher of Science* (Oxford: Clarendon Press, 1991); and Menachem Fisch and Simon Schaffer, eds., *William Whewell: A Composite Portrait* (Oxford: Clarendon Press, 1991).
19. William Whewell, "Essay Towards a First Approximation to a Map of Cotidal Lines," *Philosophical Transactions of the Royal Society of London* 123 (1833): 147–236.
20. Ibid., 148.
21. William Whewell, "Memoranda and Directions for Tide Observations," *Nautical Magazine* 2 (1833): 665.
22. Whewell, "Essay Towards a First Approximation," 163.
23. Ibid., 227.
24. For the graphical turn in Victorian science, see Thomas L. Hankins, "A 'Large and Graceful Sinuosity': John Herschel's Graphical Method," *Isis* 97 (2006): 605–633; and Reidy, *Tides of History*.
25. For Humboldt's fascination with mountains, see Aaron Sachs, *The Humboldt Current: Nineteenth-Century Exploration and the Roots of American Environmentalism* (New York: Viking, 2006).

26. Hooker to Darwin, 26 February 1854, letter 1557, in *The Correspondence of Charles Darwin*, vol. 5, ed. Frederick Burckhardt and Sydney Smith (Cambridge: Cambridge University Press, 1985–2009) (hereafter cited as the Darwin Correspondence).
27. Hooker went on several other exploratory voyages later in his career, including one to the interior of the United States. For Hooker's biogeography, see Ray Desmond, *Sir Joseph Dalton Hooker: Traveller and Plant Collector* (Woodbridge, UK: Antique Collectors Club, 1999); Janet Browne, *The Secular Ark: Studies in the History of Biogeography* (New Haven: Yale University Press, 1983); and Richard Bellon, "Joseph Hooker and the Progress of Botany, 1845–1865" (Ph.D. diss., University of Washington, 2000).
28. See the Darwin Correspondence; and Janet Browne, "The Charles Darwin–Joseph Hooker Correspondence: An Analysis of Manuscript Resources and Their Use in Biography," *Journal of the Society for the Bibliography of Natural History* 8 (1978): 351–366.
29. In addition to Darwin's and Wallace's own emphasis on Humboldt's contributions, see esp. J. D. Hooker, "On Geographical Distribution," *Report of the British Association for the Advancement of Science for the Year 1881* (London, 1882), 727–738; and, more recently, Mark V. Lomolino, Dov F. Sax, and James H. Brown, eds., *Foundations of Biogeography: Classic Papers with Commentaries* (Chicago: University of Chicago Press, 2004), 7.
30. Lomolino, Sax, and Brown, *Foundations*, 49.
31. Alexander von Humboldt, "Essay on the Geography of Plants," in *Foundations of Biogeography: Classic Papers with Commentaries*, ed. Mark V. Lomolino, Dov F. Sax, and James H. Brown (Chicago: University of Chicago Press, 2004), 52.
32. Ibid., 53.
33. Ibid., 55.
34. Darwin Correspondence, Hooker to Darwin, 3 September 1844. For a discussion of Hooker's views as they differ from Humboldt's on the relation of plant life to altitude, see esp. Bellon, "Hooker and the Progress of Botany," 181–199.
35. Darwin Correspondence, Hooker to Darwin, 28 October 1844.
36. Ibid., Hooker to Darwin, February 1845.
37. Ibid.
38. Joseph Dalton Hooker, *Himalayan Journals: Notes of a Naturalist in Bengal, the Sikkim and Nepal Himalayas, the Khasia Mountains & etc.*, 2 vols. (1855; repr., New Delhi: Today and Tomorrow's Printers and Publishers, 2005), 1:v.
39. See esp. Leonard Huxley, *Life and Letters of Joseph Dalton Hooker*, 2 vols. (London: John Murray, 1918), 1:251–264.
40. Desmond, *Sir Joseph Dalton Hooker*, 166; and Bellon, "Hooker and the Progress of Botany," 200.
41. Hooker, *Himalayan Journals*, 1:112.
42. Ibid., 1:168.
43. Peter Raby, *Bright Paradise: Victorian Scientific Travellers* (London: Chatto and Windus, 1996), esp. chap. 5.
44. Hooker, *Himalayan Journals*, 1:92.
45. Ibid., 1:325–326.
46. Ibid., 2:28.

47. L. Huxley, *Life and Letters*, 1:304.
48. Ibid., 1:325.
49. Ibid., 1:305.
50. See, for example, Raby, *Bright Paradise*; and Desmond, *Sir Joseph Dalton Hooker*.
51. Hooker, *Himalayan Journals*, 2:206.
52. Desmond, *Sir Joseph Dalton Hooker*, 166.
53. Charles Darwin, *On the Origin of Species*, ed. and introd. by J. W. Burrow (1859; repr., New York: Penguin, 1985), 367.
54. Ibid.
55. Sigmund Freud, *Civilization and Its Discontents*, trans. and ed. by James Strachey (1930; repr., London: Norton, 1961), 109–110.
56. Ibid., 44–45, 47.
57. L. Huxley, *Life and Letters*, 1:327.

Two

Lynn K. Nyhart

Emigrants and Pioneers

Moritz Wagner's "Law of Migration" in Context

This is a tale of a traveler and his baggage.

The traveler was Moritz Wagner (1813–1887), who roamed the globe from the late 1830s through the late 1850s. As a journalist and travel writer, he combined natural history, local human interest stories, and social and political commentary in a series of some half dozen books, as well as serving as a regular correspondent to the magazines and newspapers of the Cotta publishing empire. In the 1860s Wagner traversed two new boundaries: his professional career crossed over from world-traveler-cum-journalist to museum curator and professional scientist, while his intellectual surroundings shifted from a pre-Darwinian world to a fundamentally evolutionary one. Across these boundaries Wagner shouldered the intellectual baggage he had accumulated on his travels, seeking to put its contents to use in his new location.[1]

It was a hard slog, and the personal costs were high—isolation, bitterness, and ultimately suicide. But if we want to explore what it might have meant to know about the dynamics of studying biogeography on a global scale in the nineteenth century, and how that knowledge changed from the 1840s to the 1870s, we could do worse than to study Wagner. His central issues—expansion of territory, migration, isolation, and environmental determinism—were key political, social, and scientific issues attached to the global movements of his day. The form they took in his best-known work, the 1868 pamphlet *The Darwinian Theory and the Law of Migration of Organisms*, was a biological argument about the nature of evolution. But the sources of his argument, I show, lay in his broad and deep pre-Darwinian work as a travel writer who combined a study of peoples, organisms, and their environments. Even if his theory looked like one about nonhuman "nature," it was always also about people and how they made their way around the globe.

An analysis of Wagner's theory of evolution highlights a number of themes in the history of the field sciences. Most immediately, it brings our attention to questions about how naturalists scale up, generalizing from the particulars they see around them to make global claims. Evolution, whether in a Darwinian or non-Darwinian formulation, has always been a theory that sought to universalize and globalize natural events most often experienced by humans on a local level. While some evolutionists might seek to extrapolate from particular local conditions, those with the greatest authority, such as Charles Darwin and Alfred Russel Wallace, actually studied nature in different parts of the globe, and could therefore draw on geographically disparate examples to make their generalizations. So too did Wagner. This consideration leads us to a second one, about how scientific knowledge is authorized. Much historiographic attention has been given to the acquisition and appropriation of lay knowledge on the one hand, and the authority to make universal statements on the other. Because of its close relation to biogeography, evolution is one area where travelers with firsthand experience of many different locations might contend with theorists at centers of knowledge production in making claims to global-scale knowledge—though as we will see with Wagner, such experience alone rarely sufficed.[2]

Finally, unearthing the intellectual roots of Wagner's theory connects the history of field science to the history of environmental thought in some perhaps unexpected ways. Environmental historians have taken their main task to be studying the effects of humans on their environments, as well as cultural perceptions of "nature," while historians of anthropological and medical ideas have attended to past theories about the effects of climate and environment in shaping human races and their health.[3] As this essay will show, other aspects of environmental interaction also engaged thinkers of the mid-nineteenth century. Wagner was a strong environmental determinist who believed that evolution in animals, plants, and humans resulted from their interactions with their physical environments, and that the globalization of the human species was already having a profound impact on evolution (considerations that are echoed today among prognosticators on the future course of evolution).[4] His theorizing reminds us that natural history in the mid-nineteenth century could still encompass humans—and not just "primitive" ones—within its broad framing of nature. By calling attention to this side of Wagner's work, I thus hope to contribute to some broader new directions in understanding how historical actors thought about human-environment interactions.

I begin by examining Wagner's law of migration as it looked in his context of 1868, when it appeared as a contribution to the debates over Darwin's theory of evolution. I briefly consider the reaction by other scientists and Wagner's response before turning to an analysis of the enduring commitments he showed, which had their roots in his much earlier work. Territorial expansion, migration, and isolation were all themes Wagner developed in the context of analyzing human movements and the development of civilization across the globe. Their translation into the language of evolution was not unproblematic, but it was undeniably powerful.

Wagner as an Evolutionary Theorist

Wagner's Law of Migration of Organisms

In the spring of 1868, Moritz Wagner published a pamphlet titled *The Darwinian Theory and the Law of Migration of Organisms*.[5] In it, he argued that Darwin had left open a crucial gap in his theory, which he, Wagner, would fill. The gap (which many others had noticed as well) had to do with how new species actually originated. With new variations springing up all the time, how would even favorable ones not be swamped back into the larger population? What would allow a variety to develop and split off into a new species? Wagner argued that speciation required geographical isolation, which would be preceded by migration from an original area by a small group of "colonists." Without this migration and isolation, Wagner argued, evolutionary change would never take place.

Wagner's theory reinterpreted Darwin in a fundamental way, by separating the struggle for existence from natural selection and placing migration and isolation in the intervening explanatory space. For Wagner, the struggle for survival created competition among members of the same species, but the consequence was not immediately selection. Instead, population pressure caused organisms to spread out to the edges of their territory, expanding it to its natural limits (very often rivers or mountains) in an attempt to reduce competition for scarce resources, especially food.[6] A few organisms might actively move yet further, "emigrating" from their original territory across its geographical boundaries, where they would then settle a new colony in isolation from the original stock. The unaccustomed conditions would stimulate new variations, giving selection something to work on and shaping the population into a new type or species. Without migration to new conditions, the organism's internal potential for variability would not be stimulated; without isolation, new variations would be unable to separate themselves off from the population as a whole. By interpolating migration and isolation between the struggle for existence and selection, Wagner radically reinterpreted the relationship between two of Darwin's key elements.

In presenting his theory, Wagner entered into the ongoing controversy among experts over the weaknesses of Darwin's theory and how to overcome them. Wagner's theory of evolution allowed him not only to counter the problem of swamping, he thought, but also to overcome several other difficulties as well. It answered a major objection of Darwin's German translator Heinrich Georg Bronn: by the incrementalist logic of Darwin's theory, intermediate forms should be found everywhere, but they are not. Migration and isolation removed this objection, in Wagner's view, because these mechanisms were not exactly gradualistic. Instead, the new colony functioned as a founder population (to use a modern term), and its evolutionary potential would be rapidly unleashed to form a new race—a potential new species. A second objection was that Darwin's theory would seem to leave the persistence of lower forms unexplained. Why didn't all organisms improve over time? Wagner's answer: those that didn't improve were organisms with continuous territory that was either so limited that none could escape, or so broad (as in the case of cosmopolitan organisms) that they had nowhere to go that would be separate and isolated from the stem population. Without a new location, variability would not be stimulated. The same reasoning also answered the ever-reappearing objection of the similarity of ancient Egyptian forms to modern-day ones. The Egyptian ibis and crocodile were confined to a fixed territory, with no possibility of leaving, hence they were not subject to change.

Wagner argued that the same processes operated on humans as well: "Individual human pairs must have migrated, often far beyond the furthest limits of the distribution area of their crude race-mates, driven by the wish to improve their conditions of life." With complete isolation from their former population, such originating pairs might produce a new variety or prospective race. Wagner noted that the high mountain regions were always considered the cradle of civilization among different cultural traditions—the Himalaya and other central Asian mountains for the Mongolian race, the Ararat group and Armenian Taurus for the Semites, the mountains from the Caucasus to the Hindu Kush for the Aryans. Likewise, the Atlas Mountains were the legendary origin of the peoples of northwestern Africa, the East African highlands for the Abyssinians and Nubians, and the mountains of Mexico and Peru for the ancient cultures there. For good reason, Wagner thought: these mountains provided the necessary isolation that would stimulate new racial development from founding pairs, as well as the hard living conditions that would enable only the strongest and most intelligent of their offspring to survive. Mountains drove human evolution just as they pushed animals and plants to evolve. Populations that lived in the open spaces of the plains stagnated and did not progress.[7]

Wagner's theory led him to some startling conclusions. One consequence was that since humans had populated virtually the whole world, evolution was now grinding to a halt. Human expansion had radically limited the ability of other organisms to migrate freely from their original locations, and so the only real possibility for further evolution in plants and animals would come from artificial selection and hybridization at human hands. By the same token, no new human races would evolve, since complete isolation of humans from one another was no longer possible. The only future evolution of man would come in two ways: through hybridization of two major races, which might create a stable new race; and through cultural forms of isolation and mutual antipathy organized around "differences in estate, religion, [and] nationality."[8] The consequences were somewhat ominous, as Wagner, quoting Aristotle, declared that the most fundamental principle of nature is change.[9] Holding his best card for last, his final sentence quoted not Darwin but a hero of biogeography who loomed still greater for Germans of Wagner's generation, Alexander von Humboldt: "The spatial separation [Abgrenzung] of form, a necessary consequence of migration, is the cause of its typical difference."[10]

Wagner rested his claims to authority to make generalizations about evolution and biogeography on his experience as a global, Humboldtian-style traveler who sought world-scale generalizations deriving from his own firsthand experience of locations across Europe, northern Africa, Asia Minor, southern Russia, and North, Central, and South America. His pages are filled with biogeographical examples from these places, which together reinforced the message that closely related species often lived nearby one another but were separated by a significant geographical barrier, such as a river, mountain, or ocean.[11] Wagner had noted these biogeographical patterns long since, but Darwin had offered a new explanation for them—which Wagner, based on his own experience and knowledge, found incomplete. To modern readers, Wagner's lengthy examples of distribution look like mere expansions on a point that Darwin had already made fully in his two chapters on geographical distribution in *On the Origin of Species*.[12] But through these examples, I suggest, Wagner was actually seeking to establish to the scientific community both his own priority of discovery of these biogeographical patterns (though not their explanation) and his authority to oppose Darwin's interpretation of them. Wagner did say that he had previously found these patterns "puzzling" and inexplicable until he read Darwin, though he had thought much about them. But while he "did recognize a certain coherent connection [Zusammenhang]" that Darwin brought to the phenomena, he did not credit Darwin with changing his perspective on these patterns. He simply could not swallow Darwin's explanation of distribution, "even after repeated careful reading" of Darwin's work.[13] Just as Darwin's German translator Bronn believed that his career-long experience as a paleontologist licensed him both to translate *Origin* and criticize it in within the same covers, Wagner claimed his own authority to modify Darwin's theory on the basis of his career as a traveling naturalist.[14]

The Zoologists' Response: Wagner as a Layman

The authority of the traveler-naturalist did not carry very far within the German zoological community of the late 1860s and 1870s. In particular, August Weismann and Ernst Haeckel, leaders of a new generation of up-and-coming professional zoologists, denigrated Wagner's theory as insufficiently Darwinian—indeed, as insufficiently scientific.

Haeckel read Wagner's pamphlet as he was readying the second edition of his popularly aimed *Natürliche Schöpfungsgeschichte* (Natural history of creation) for publication, and he incorporated a basic critique into its pages. Wagner seemed only to incorporate sexually reproducing beings into his theory, Haeckel wrote, and omitted the important role played by asexual reproduction in many organisms, such as marine invertebrates. Haeckel conceded that migration and isolation could speed the production of new species in sexually reproducing organisms, but he could not agree that it was required. Although he said that Wagner was completely wrong in this claim, he was otherwise quite generous about the significance of Wagner's evidence for migration and isolation for the production of new species, referring readers interested in the geographical distribution of plants and animals to Wagner's pamphlet along with works by Darwin and Wallace, and (temporarily) retitling the theory of evolution "the theory of selection and migration." Once it was restricted to sexually reproducing organisms, Haeckel had little more to say against Wagner's theory. Instead of mounting further specific arguments, he referred his readers to a lecture given by Weismann, who had argued that speciation certainly was possible without geographic isolation.[15]

Weismann himself was more aggressive. His critique, like Haeckel's inserted into a previously completed work, bluntly denied the overriding significance Wagner accorded to migration. Natural selection was primary, and isolation was a consequence of it, not the other way around, as Wagner would have it. Moreover, Weismann thought there was evidence that new races could form without geographical isolation, and conversely, that migration and isolation might both occur without the formation of new species.[16] Both men found geographical isolation a useful but not necessary aspect of species formation.

In 1870 Wagner replied. In response to Weismann, he politely acknowledged the latter's criticism, but declined to accept his interpretation of the evidence. In response to Haeckel, he modified his claims in two ways. First, he acknowledged the need to limit his theory to sexually reproducing organisms, although he thought that scientists still knew little enough about asexual marine invertebrates not to rule out the possibility of something like his theory of isolation turning out to be true for them as well. Second, he became a Lamarckian. Although there is no direct evidence that Haeckel's work was what stimulated Wagner's conversion, it seems plausible, for Haeckel called new attention to it in his *Generelle Morphologie* (General morphology) and *Natürliche Schöpfungsgeschichte*. Wagner clearly had not read Lamarck until after 1868, the year the latter book was published. In the French author Wagner found a new source of confidence that adaptation to new environments stimulated the variations that would create new species. Isolation still was required to prevent those variations from being swamped back into the original type, but now he thought that, given the power of Lamarckian adaptation, his theory didn't actually require selection at all.[17]

This was too much for Weismann. In 1872, he produced his own pamphlet, which skewered Wagner for misunderstanding Darwin, drawing illogical conclusions, and ignoring evidence that countered Wagner's own theory.[18] Wagner responded with new defenses in a long series of articles in Cotta's popular geographical magazine *Das Ausland* (Abroad). But Weismann was done with him: Wagner was no longer worth his time.

A clue to both Weismann's initial attention and his later inattention comes in the foreword of his 1872 pamphlet. Although Wagner's theory was undoubtedly wrong, and poorly argued to boot, Weismann thought, it was still worth rebutting, "especially in an area that threatens to become a domain of dilettantism"—a not-so-subtle jab at Wagner himself.[19] It may also have signaled Weismann's disapproval of the attention Wagner was getting. Certainly Haeckel was dismayed, as he wrote to Weismann in 1871, while the latter was working on his response to Wagner: "It is remarkable how little sound criticism exists with respect to such 'theories.' Not only in Munich [Wagner's home], but also in many other places, this 'theory of migration' is set side-by-side with or even above the selection theory of Darwin." Haeckel hoped that he would often find Weismann by his side as a "brave fellow combatant" in the struggles over the theory of evolution. Weismann replied that refuting Wagner was "easy," but that coming to a positive understanding of the effect of isolation on transformation was more complicated.[20]

Here Haeckel and Weismann stood shoulder to shoulder against the "dilettantism" of a man who seemed to come out of nowhere. "Do you happen to know M. Wagner personally," Weismann wrote Haeckel in a postscript, "and can you tell me what sort of customer he actually is?"[21] No reply from Haeckel has been preserved. What they might have known about him was that he was a curator of ethnology, not of natural history, at the Bavarian state collections in Munich; and that he had no proper Ph.D., just an honorary degree from the University of Göttingen, probably wrangled through the offices of his much older brother, the Göttingen professor of anatomy and physiology Rudolf Wagner. If they had kept abreast of academic politics in Bavaria, they might have heard rumors that he had failed in his bid for a professorship in Munich because those seeking to upgrade Bavarian science considered his travel writings old-fashioned and unscientific, and that the creation of the state ethnological curatorship for him was a consolation prize.[22] The problem wasn't exactly that he wrote for magazines: in the early 1870s both Weismann and Haeckel were still eager to engage a wider public, and neither shied away from writing newspaper and popular magazine articles or speaking to public audiences. The problem was that by their standards (like those of their Bavarian counterparts) he was an amateur, and they found it annoying to have to spend their time explaining why amateurs like Wagner were wrong.

Historians of biology have tended to sympathize with the professionals. Fred Churchill reflected Weismann's perspective in calling Wagner's argument "amateurish."[23] More dramatically, Frank Sulloway has described Wagner's commitment to the creative powers of isolation combined with Lamarckian adaptation as "unconvincing and even fantastic," and "almost mystical."[24] But Sulloway's characterization only begs the question of why Wagner found his own arguments so persuasive. What was it about migration and isolation that led him to place these factors so far ahead of others in his theorizing about evolution?

While Wagner's increasing sense of embattlement may certainly have contributed to his raising the rhetorical stakes over the course of the 1870s, historians have missed the larger framework in which he developed his evolutionary theory. This was not, fundamentally, a biological argument as understood by professional biologists like Weismann and Haeckel. It derived instead from (and to some extent remained within) a geographical-environmental argument about the interactions of plants, animals, and humans—especially German emigrants—with the physical features of the globe, which Wagner had developed as a travel writer in the 1840s and 1850s. According to Wagner these interactions, expressed through territorial expansion, military conquest, and migration in its several forms, formed the basis of the history of civilization. It was not a large step to expand this notion to make it the basis of the history of nature.

Wagner on Travel

Wagner's travel books charted three particular kinds of human movement: his own, as he journeyed through foreign lands, talking to people he met, often picking up one or a few traveling companions for a portion of the trip; the imperial expansion of the major political powers, especially the French, the Austrians, and the Russians; and the emigration of Germans, mainly in small groups, eastward as far as the Caucasus and westward to the Americas. Although his own journeys as a traveler provided both the narrative structure for these books and the occasions for remarking on his experience of new places, his observations about imperial expansion and colony building reveal more about his ideas on the dynamics among humans and their living and physical environments. Let us consider each aspect of human movement in turn—Wagner's own travels, imperial expansion, and German emigration—to view their relations to the natural and physical environment as Wagner presented them.

Wagner's Travels

Wagner's earliest journeying took him to northern Africa, where from 1836 to 1838 he reported on the French efforts to colonize Algeria. From 1843 to 1846 he traveled the southern side of the Black Sea and across to the Caspian, observing life, politics, and nature in Turkey, Armenia, the area on the southeastern shore of the Black Sea known as Colchis, the southern Caucasus, and Persia, with special attention to German settlements in Colchis and the southern Caucasus. He then returned home to southwestern Germany, where he wrote up his findings in four books published between 1848 and 1852, while also reporting on the revolutionary activities of 1847–1848. Disillusioned at the failure of the revolution, he sought solace in further travel—this time across the Atlantic to North America, where hundreds of thousands of Germans were settling. Long interested in the patterns and possibilities of German outmigration, his works on North America and Costa Rica (products of his travels with Carl Scherzer from 1852 to 1854) served in part as manuals of practical advice to Germans thinking of emigrating.[25]

Wagner's books in the 1840s and 1850s concentrated primarily on stories about people, politics, history, and their relation to the landscape, but natural history came into his work as a traveler in various ways. To begin with, he was a serious collector of the fauna of the lands he roamed. Indeed, he had taken up journalism as a foreign correspondent precisely because it would afford him the chance to travel, collect specimens, and study nature, since family economic circumstances prevented him from obtaining a university degree.[26]

His travel narratives occasionally referred to his collecting activities, and some also included separate appendices on the natural history of particular locales, which included lists of species and remarks on their distribution.[27] In comparison to the importance of his physical collections, however, the amount of technical natural history—descriptions of organisms and lists of species—seems relatively understated. In the case of his travels to western Asia, this may have had to do with his intention to write a separate monograph on the topic—a hope shattered by the contraction of the publishing industry after 1848.[28] More prominent were his vivid descriptions of the landscapes through which he traveled, which were woven into his narratives.

Both the specific discussions of faunal distribution and his attention to landscape betray a strong debt to Alexander von Humboldt, one of his models and patrons. Wagner's close attention to the altitude at which different forms could be found, to whether those forms were the same as those at the same latitudes in Europe, and to whether they might be considered "characteristic" of a particular location, suggest an effort to apply to animals Humboldt's famous approach linking latitude, altitude, and plant distribution. More broadly, Humboldt's *Ansichten der Natur* had treated landscape simultaneously as an aesthetic and scientific object, and Wagner's writing similarly combined warm and romantic descriptive writing with attention to the physical properties of the landscape in accounting for distribution, such as the effects of mountain ranges as barriers and river valleys as conduits for species expansion. Indeed, the dynamics of geographical distribution—the extension of territory, the processes of migration and diffusion and the barriers to those processes, the effects of different environments on organisms (including humans)—were what most interested him about nature. When writing about humans, he blended this environmentally deterministic focus with political and ethnographic analysis to account for human movements across the globe.

Imperial Expansion

A major theme of Wagner's writing concerned the spread of civilization through imperial expansion. In fact, his 1868 evolutionary pamphlet was not the first time he had posited a "law of migration"—a law by that same name appeared in his 1854 account of his travels to North America. But this one did not focus on animals or plants; rather, it concerned the spread of civilization. He wrote: "Civilization generally *must*, it would appear, somewhere make conquests in order to survive [leben]. The existence of a 'law of migration,' according to which culture and civilization [Bildung] *must* spread themselves over the entire earth, is a truth among astute researchers as recognized as the course of the constellations and the circulation of the blood in the human body."[29] Territorial gain was the main manifestation of the spread of civilization—and of civilized peoples and their governments. Here geography played a key mediating role, for just as the topography and physical attributes of a region dictated the distribution of plants and animals, so too did these features structure the possibilities for military conquest and nationhood.

Throughout Wagner's travel writings, examples can be found of geography determining the shape of nations and empires. Thus the German and Italian states were small and multiple because of their dividing terrain, which isolated "tribes and states" and hindered efforts at unification; a conqueror could easily overtake Poland because it was flat and offered ready access to military control.[30] Comparing the efforts of the French to conquer Algeria with those of the Russians in the Caucasus, he noted that the Caucasus formed an isthmus between the Black and Caspian Seas. Although its native dwellers might find refuge in its forested mountains, which impeded the Russian cavalry's movement, ultimately the topography would allow conquest, because the Russian army could control both ends of the isthmus. The Atlas Mountains of northwest Africa, by contrast, led into the desert—a vast expanse impossible for Europeans to surround.[31]

Topography alone did not determine the course of history: as Wagner's discussion of the Russian cavalry hints, military strategies and organization were important, and broader cultural factors also played a role. Wagner noted that if the Caucasus dwellers were as fanatically religious and united in culture and language as the North African Berbers, the Russians would have a much harder time conquering them. In further contrast to the French, the Russians also knew how to use agricultural settlement to their advantage, transplanting sympathetic peoples who would support them in their expansionist quest. As long as the French effort to conquer Algeria remained military, and not one of European settlement, Wagner argued, they would never control the territory.[32]

Wagner viewed the history and future of the United States through a similar geographical lens. In an assessment that placed physical geography first and culture as an auxiliary shaper of history, he declared that the topography of North America dictated that it would ultimately be a single national unit. (The explicit context here in 1854 was the increasing threat that the southern United States might split into a separate country). The most striking physical feature of North America, according to Wagner, was its readily navigable waterways and relatively low mountains, which combined to allow ready transportation and communication across the vast plain between the mountain ranges near the East and West coasts, which themselves were easily traversable. Under such circumstances, he argued, economic interests would dictate unity, and political and social unity would inevitably follow, through "the rapid transformation and merging" of previously distinct national, linguistic, and racial identities, which would either be assimilated into the Anglo-American identity or die out: "It was the will of the Creator, who, as he formed the three-dimensional outlines of the new continent in contrast to the structure of the eastern half of the world, held out the promise of victory to only *one* race, the most powerful among all, and granted only *one* nation existence here." The structure of the land dictated that all other races and nations would die out, and that only this amalgamated one would survive.[33] (Recall that this was written in 1854, five years before Darwin published *On the Origin of Species*!)

Whether the united country that resulted would be a democracy was less certain. Wagner—a committed democrat who had been profoundly disappointed by the failures of 1848—argued that the commitment to democracy and self-government of the individual American states would succeed only as long as "fallow and uninhabited wilderness lies before them and the colonization and populating of the wilderness is the main task of these states." Nearly forty years before Frederick Jackson Turner broached the same argument, Wagner posited that once the population could no longer expand into the frontier, a profound change in the nation would likely ensue. The open topography of the land from coast to coast would once again push the economy toward unity, and therefore would tend toward a single, strong, national government. A new aristocracy would emerge, he predicted, as a filled-in country developed economic and social inequalities—perhaps unlike the European aristocracy, deriving not from blood but rather from merit, but a ruling caste nevertheless. Even monarchical rule was not an impossibility, at least based on the continent's landform. If democratic rule persisted in North America, Wagner argued, it was not because this was a "natural" state dictated by geography, but because the continent had been settled by political and religious refugees, "freedom-thirsting people" who demanded political equality and independence. Geographical conditions would not help them out.[34]

Elsewhere in the volumes on North America, Wagner contemplated the effects of the territorial expansion of civilization on the native peoples. In line with the theory of migration that he had posited near the beginning of the book, he considered the victory of civilization to be inevitable, but he saw two ways that might happen, depending on the actions of both civilized and primitive groups. If the American Indians allowed themselves to be assimilated and civilized, they would persist, and join into the strong, new national community that was being formed. This was also the object of the truly civilized pioneers of civilization. The Indians would of course resist, but if appropriate care were taken, they would survive. But the unpleasant side of American society was its restless lust after money, its haste and lack of human sympathy, which would drive the conquerors to trample over the Indians rather than assimilate them, leaving as their legacy only the many place-names they had bequeathed to the land.[35]

Here we see Wagner's characteristic concern with a three-way balance among human movement—especially territorial expansion and the interactions of populations or tribes—geographical topography, and the development of political and social forms. The story of civilization was the story of the interactions among these factors, and Wagner saw himself as its interpreter, not only for past and present, but also for prognosticating the future.

Wagner's experience as an observer of imperial expansion and the struggles entailed in it shaped his evolutionary theory in profound ways. First, of course, there is his commitment to the primacy of territorial expansion. His 1868 law of migration expanded this commitment from the history of civilization to the processes of nature writ large, while effacing its origins in his earlier theory of human history. But his language remained heavily laden with metaphors drawn from this human imperial history, and especially from his travels in North and Central America. In a brief but significant passage in the 1868 work, he wrote that the plant communities of the jungle are in a struggle for territory against those of the savanna: "Here certain tree forms have originated . . . that play the role of the penetrating 'pioneers' among the emigrants from the forest, which have adapted themselves to the dryness and the light-stimulation of the savanna, and that grow only at the outermost edges of the forest, which therefore must advance against the savanna as the boundary of the forest expands." An accompanying footnote remarks on the "continuous struggle of the forest against the savanna" in tropical America, which the forest would win if only the "Indians and colonists" did not keep it back by periodic burning.[36] While easily read in Darwinian (or quasi-Darwinian) terms, this passage represents the earliest application I have seen of the term "pioneer" to plant species, in exactly the same ways that later ecologists would use it. Wagner's observations of German pioneers in America clearing land for agriculture, combined with his analysis of the front line of imperial expansion, lends a new specificity to our understanding of his analogizing from humans to plants. It was not just a theory for him: he had witnessed both the savannah-jungle boundary and the farmers at work clearing (and invading) the plains, on the same journey where he had observed the larger process of civilized peoples taking over the lands of the nomadic American Indians (not dissimilar to the efforts of the Russians to colonize the Caucasus or the French to take over northwest Africa, but more successful). It was all of a piece for him.

Free Settlements

If nation and empire building and their dependence on geography constituted a major theme in Wagner's travel writings that would carry over into his later evolutionary writings, his analysis of smaller-scale migrations, in the form of free settlements, also held a key place. In his 1850 book *Reise nach Kolchis und nach den deutschen Colonien jenseits des Kaukasus* (Journey to

Colchis and to the German colonies beyond the Caucasus), Wagner was centrally concerned with German émigrés (especially religious émigrés) who had settled in the Caucasus and Transcaucasus. These German "colonies" had been established in the late 1810s, in the wake of the Napoleonic Wars, by a combination of migrants. Some had left southwestern Germany for economic reasons, seeking land that was more fertile than the worn-out soil of their native Swabia (a region that ran across the political borders of Baden, Württemberg, and Bavaria). Others had left for religious reasons, including those who sought fresh spaces where they would be free to worship without religious oppression and those who sought to be closer to Jerusalem when the soon-to-arrive Day of Judgment would come. (Many German millenarians viewed the Napoleonic Wars as a sign of the Last Days.) Southern Georgia and the Caucasus beckoned them, made more attractive by the welcoming policies of the Russian central government, which invited settlements by "civilized" Germans in the southern reaches of Russia's expanding empire, in the hopes that they would have an uplifting effect on the native tribal populations who surrounded them.[37]

A hallmark of these German settlements was their separateness. When Wagner visited Katharinenfeld (now Bolnisi, Georgia), Elisabeththal, and five other nearby German settlements, he found all together over six hundred families comprising three thousand German individuals.[38] They were not integrated into the local economy or culture: the butter and vegetables they produced, though very fine, were not purchased by locals, who considered the products foreign and too expensive. They had reconstructed "a piece of Germany" with their clocks and easy chairs, their tidy vegetable gardens, and their neatly sown fields, adopting neither the elegance of oriental style nor the relaxed relationship to the wild landscape of those who surrounded them.[39]

They were vulnerable as well. In Katharinenfeld in the mid-1840s, the memory still lingered of savage attacks in 1826 by a band of Tartars, who had killed thirty people, kidnapped sixty-five, and stolen all the portable goods. The village was rebuilt with loans from the Russian government, but the people still shuddered at the memories. Religious separatism remained strong. While Wagner was visiting the village, a comet passed and a woman had a vision of the Last Days. Millenarians from the surrounding German settlements stopped their work and prepared to travel to Jerusalem. The local Russian governor-general vowed to prevent them from leaving—they could not leave before paying off their (considerable) debt to the government—but the pilgrims were undeterred: "What could the Cossacks do against the angelic horde? What could the prohibition decreed by a royal general do against the command decreed by the Lord God?" The pilgrims prepared; the army massed. When the day to leave arrived, no bolt came from the blue, no miracle occurred, and the army was still there. "The poor people rubbed their eyes, as if they were awakening from a strange dream," and went back home—now with the further cost of calling up the army to pay for.[40]

Wagner's description of his experiences with the German colonists highlighted the small and fragile nature of their colonies—only through determination, German hard work, and luck did they survive, in the face of unexpected hardship, poor weather, lack of help from civilized locals, and attacks from uncivilized ones. Their faith helped some of them (though it is clear that Wagner thought many were simply crazy). Culturally, too, they were profoundly isolated. What was perhaps most striking to him was their lack of interest in news Wagner had brought from the Swabian homeland they had left. Although wedded to their memories of their past, they cared not at all for the present and future of Germany.[41] In Wagner's view, though they had not mixed with the locals, they were no longer true Germans in sentiment. They were a new entity, a group whose cultural, economic, and personal peculiarities had driven them from home in the first place and formed the basis of their new colony. This was the model of isolation that would most obviously haunt Wagner's evolutionary theory of migration.

Wagner's Theory Revisited

With Wagner's long history as a traveler and analyst of the history of human migration (especially German migration) in mind, his commitments to the primacy of territoriality, migration, and isolation in evolution become clearer. Once we know about his understanding of French and Russian efforts at imperial expansion, his special attention to topography, his tracing of German colonists in the Caucasus, and his analysis of German farmers settling the North American frontiers, we find new meaning in central aspects of his theory and as well as its sidelights. Migration for territorial expansion had long been a "law" for Wagner. Formerly a law of civilization and the expansion of its powers, by 1868 it was a law of nature. But migration at a smaller scale was also a way of avoiding competition, of escaping the pressures of an intense struggle for existence. Swabians had known that for a long time; German democrats had learned the lesson more recently; evolutionists were just beginning to see it. Is it any wonder that Wagner's biological theory placed migration and isolation ahead of "selection" (which, after all, had that difficult problem of personifying nature as "selector")?

What else might we learn from Wagner's tale? First, it affords us an example of how the field-based study of humans could intersect with analysis of the rest of nature in the nineteenth century. Although it is easy to view Wagner's intellectual path as an extrapolation from observing humans in a pre-Darwinian context to making claims about all of nature in an evolutionary context, the situation is more complicated. Both the travel writings and the evolutionary writings dealt with nature as a whole, and viewed humans as subject to nature and acting in interaction with physical nature, just as animals and plants were. In generalizing about migration and isolation, Wagner seems to have merged his considerations about humans more fully with those about plants and animals, to come up with a unified theory (one that was different from Darwin's). It seems quite clearly a theory based on the naturalization of imperial expansion and conflict, but his previous theory of empire itself was also nature-based, deeply dependent as it was on physical geography and on drawing analogies from the plant and animal world to the human one. It reminds us that we would do well not to assume that the naturalization of empire was a post-Darwinian phenomenon.

Nor, second, was the naturalized empire necessarily a British one. When historians think about the history of colony building, especially in the expansionist nineteenth century, we have tended to view colonies as outposts of empires, drawing most often on a British model of imperial conquest. But as Wagner viewed the German form of free colony settlement, it was not so much an expansion of an empire to a new realm as an escape—from overpopulation, or from religious or political oppression. Germans settled in "colonies" across the world, but they did not generally see themselves as building an empire before the nation itself was united in 1871.[42] German settlements, in fact, were recognized even in the mid-nineteenth century as exemplifying a different model of colony making than that of Britain.[43] Only as German emigration shifted from a phenomenon of sporadic bursts to a huge, steady outflow beginning at mid-century, and as the German states finally came together into the second empire, did Germans come to see a nation-based empire as their model too. (Although Wagner witnessed this shift, he did not incorporate it into his thinking. The naturalization of a larger-scale model of migration and colonization tied to imperial expansion would be taken up by another traveler-journalist-geographer, Wagner's younger friend and protégé Friedrich Ratzel.)[44] As Wagner's theory of evolution shows, the model of small, isolated settler colonies offered a different way of thinking about migration and distribution than that presented by the nineteenth-century British Empire.

Finally, however, we also need to recall that Wagner's theory, though it belonged to a German traveler, was not adopted by most other German evolutionists. It was not "natural" for all German scientists to adopt a migration-isolation model of evolution, because what was so central to Wagner's experience was not so to most other students of evolution. He was a world traveler, a field collector, and a prognosticator. To the extent that German university zoologists of the early 1870s did any field science (which was less and less as the prestige of the lab increased), they studied at the seashore (preferably around the Mediterranean), or perhaps at local lakes or German forests and fields; the period from the mid-1860s to the end of the 1870s was not a time of great state-sponsored natural history expeditions. Zoological professionalization and nation building in Germany combined to turn the evolutionist's eye inward, toward the lab and the cell. Field collecting did take place, to be sure—but it was becoming identified as a realm of dilettantes and professional collectors, not those who did "real" science themselves. And no professional zoologist worth his salt spoke about the state of the world or its future.

Wagner persisted in writing his most important theoretical works about evolution in high-end, semi-popular magazines such as the Cotta publication *Ausland*, a magazine of geography and world news, and *Kosmos*, a popular evolution-promoting magazine. This marked him yet more as an amateur, someone who appealed to a broad audience because he was unwilling (or unable) to conform to the standards and practices of professional science as these were coming to be defined. This was a scientist of the 1840s, perhaps, working in the mold of an Alexander von Humboldt, but by 1870 Humboldt had been dead for over a decade and his name co-opted by popularizers.[45] Old-style geography and ethnology were outmoded; a new style, which would gain legitimacy with the growing German empire, was yet to emerge. The baggage Wagner carried with him marked him as a man from an earlier era, a world traveler lost in the professionalized German zoology of the 1870s.

Notes

1. It would be interesting to study his actual baggage, especially the substantial collections he made during his travels, but currently not enough information is available to pursue this.
2. Historians of medicine were among the earliest to call attention to the role of lay knowledge and its importance in shaping and contesting the knowledge of experts. See Roy Porter, ed., *Patients and Practitioners: Lay Perceptions of Medicine in Pre-industrial Society* (Cambridge: Cambridge University Press, 1985). For recent references specific to lay knowledge and the history of the field sciences, see Henrika Kuklick and Robert E. Kohler, eds., *Science in the Field, Osiris* 11 (1996); Daniel W. Schneider, "Local Knowledge, Environmental Politics, and the Founding of Ecology in the United States: Stephen Forbes and 'The Lake as a Microcosm' (1887)," *Isis* 91 (2000): 681–705; and Florian Charvolin, André Micoud, and Lynn K. Nyhart, eds., *Des sciences citoyennes? La question de l'amateur dans les sciences naturalistes* (La Tour d'Aigues, France: Editions de l'Aube, 2007). On travelers' knowledge, see Stuart McCook, "'It May Be Truth, but It Is Not Evidence': Paul du Chaillu and the Legitimation of Evidence in the Field Sciences," *Osiris* 11 (1996): 177–197.

3. Environmental history is a large and rapidly growing field of historical inquiry; I have continued to find Bill Cronon's work an invaluable source of insight and inspiration. See William Cronon, ed., *Uncommon Ground: Rethinking the Human Place in Nature* (New York: Norton, 1996). On the history of medical geography, see Nicolaas Rupke, ed., *Medical Geography in Historical Perspective* (London: Wellcome Trust Centre, 2000); and Frank Barrett, *Disease and Geography: The History of an Idea* (Toronto: Atkinson College, Department of Geography, 2000). For a perspective that explicitly seeks to integrate environmental history with the history of health and disease, see Gregg Mitman, Michelle Murphy, and Christopher Sellers, eds., *Landscapes of Exposure: Knowledge and Illness in Modern Environments, Osiris* 19 (2004).
4. Donald Waller, "Reshaping the Tree of Life: Human Impacts on the Future of Evolution," presentation at Darwin Day, University of Wisconsin, Madison, February 9, 2008.
5. Moritz Wagner, *Die Darwin'sche Theorie und das Migrationsgesetz der Organismen (1868)*, in *Die Entstehung der Arten durch räumliche Sonderung: Gesammelte Aufsätze*, ed. Moriz Wagner (Basel, Switzerland: Benno Schwabe, 1889), 47–97. This was an expansion of a paper he had read at the Bavarian Academy of Sciences on March 2, 1868. Pagination is from the reprinted version in Wagner's collected essays, cited above. The title's English translation and all other unattributed translations are my own.
6. This idea would be elaborated much further by Wagner's student Friedrich Ratzel in his concept of the Lebensraum or "living space." Work on Ratzel and the Lebensraum concept tends to play down or even ignore Ratzel's debt to Wagner, which was substantial. See, for example, Woodruff D. Smith, "Friedrich Ratzel and the Origins of Lebensraum," *German Studies Review* 3 (1980): 51–68.
7. Wagner, *Die Darwin'sche Theorie*, 79–80, on p. 79.
8. Ibid., 86.
9. Wagner's "end of evolution" scenario, like his model of individual human pairs as the founders of races, seems to be a naturalistic interpretation of Judeo-Christian assumptions, though Wagner himself was not religious. On eschatology and the end of history in nineteenth-century earth science, see Nicolaas Rupke, "'The End of History' in the Early Picturing of Geological Time," *History of Science* 36 (1998): 61–90.
10. Wagner, *Die Darwin'sche Theorie*, 97.
11. Ibid., 56, 61. Wagner used the term "vicariant" (*vikarirende*; elsewhere, *vikarierende*) for such species, which "substitute" for one another (like a vicar for a priest) in different geographical locations that are ecologically similar. A history of ideas of vicariance from Wagner's time to the new "vicariance biogeography" of the 1970s and 1980s would be an interesting project.
12. Frank Sulloway, "Geographic Isolation in Darwin's Thinking: The Vicissitudes of a Crucial Idea," *Studies in History of Biology* 3 (1979): 23–65. Interestingly, Wagner never claimed that Darwin had led him to his new, evolutionary interpretation of the biogeographical facts. It was his eponymous nephew and biographer who made that claim for him, after his death (in Wagner, *Die Entstehung der Arten*), and whom Sulloway cites (p. 49) as evidence of the senior Wagner's debt to Darwin.
13. Wagner, *Die Darwin'sche Theorie*, 52.

14. For the best interpretation to date of Bronn and his approach to *Origin*, see Sander Gliboff, *Translation and Transformation: The Origins of German Darwinism* (Cambridge: MIT Press, 2008).
15. Ernst Haeckel, *Natürliche Schöpfungsgeschichte*, 2nd ed. (Berlin: G. Reimer, 1870), 326–332, on p. 332.
16. August Weismann, *Ueber die Berechtigung der Darwin'schen Theorie: Ein Akademischer Vortrag gehalten am 8. Juli 1868 in der Aula der Universitat zu Breiburg im Breisgau* (Leipzig, Germany: W. Engelmann, 1868), 32–36.
17. Moritz Wagner, *Ueber den Einfluss der geographischen Isolierung und Kolonienbildung auf die morphologischen Veränderungen der Organismen: Vortrag in der Sitzung der k. bayer. Akademie der Wissenschaften vom 2. Jul 1870*, in Wagner, *Die Entstehung der Arten*, 101–116.
18. August Weismann, *Ueber den Einfluss der Isolirung auf die Artbildung* (Leipzig, Germany: W. Engelmann, 1872).
19. Ibid., iv.
20. Haeckel to Weismann, 3 May 1871; Weismann to Haeckel, 31 May 1871; and Haeckel to Weismann, 18 January 1872, in Georg Uschmann and Bernhard Hassenstein, "Der Briefwechsel zwischen Ernst Haeckel und August Weismann," in *Kleine Festgabe aus Anlaß der hundertjährigen Wiederkehr der Gründung des Zoologischen Institutes der Friedrich-Schiller-Universität Jena im Jahre 1865 durch Ernst Haeckel*, ed. Manfred Gersch (Jena, Germany: Friedrich-Schiller-Universitat Jena, 1965), 28–30, 31–32.
21. Weismann to Haeckel, 31 May 1871, in ibid.
22. Wolfgang J. Smolka, *Völkerkunde in München: Voraussetzungen, Möglichkeiten, und Entwicklungslinien ihrer Institutionalisierung (ca. 1850–1933)* (Berlin: Duncker and Humblot, 1994), 61–95.
23. Frederick B. Churchill, "August Weismann, a Developmental Evolutionist," in *Ausgewählte Briefe und Dokumente: Selected Letters and Documents*, ed. August Weismann, F. B. Churchill, and Helmut Risler (Freiburg im Breisgau, Germany: Universitätsbibliothek, 1999), 2:786.
24. Sulloway, "Geographic Isolation," 55, 58.
25. Moritz Wagner, *Reisen in der Regentschaft Algier in den Jahren 1836, 1837, und 1838*, 3 vols. (Leipzig, Germany: Voss, 1841); Wagner, *Der Kaukasus und das Land der Kosaken in den Jahren 1843–1846*, 2 vols. (Leipzig, Germany: Arnoldi, 1848); Wagner, *Reise nach dem Ararat und dem Hochland Armenien* (Stuttgart, Germany: Cotta, 1848); Wagner, *Reise nach Kolchis und nach den deutschen Colonien jenseits des Kaukasus: Mit Beiträgen zur Völkerkunde und Naturgeschichte Transkaukasiens* (Leipzig, Germany: Arnoldi, 1850); Wagner, *Reise nach Persien und dem Lande der Kurden*, 2 vols. (Leipzig, Germany: Arnoldi, 1852); Wagner and Karl Scherzer, *Reisen in Nordamerika in den Jahren 1852 und 1853*, 3 vols. (Leipzig, Germany: Arnoldi, 1854); and Wagner and Scherzer, *Die Republik Costa Rica in Central-Amerika mit besonderer Berücksichtigung der Naturverhältnisse und der Frage der deutschen Auswanderung und Colonisation: Reisestudien und Skizzen aus den Jahren 1853 und 1854* (Leipzig, Germany: Arnoldi, 1856). The authorship of the separate sections of *Reisen in Nordamerika* is identified by initials. Biographical information here and below is drawn from Smolka, *Völkerkunde in München*; and Friedrich Ratzel, "Wagner, Moritz," in *Allgemeine Deutsche Biographie*, vol. 40

(Munich: Bayerische Akademie der Wissenschaften: 1896), 532–543. Ratzel notes (pp. 535–536) that the natural history part of Wagner's Costa Rica researches (Wagner, *Naturwissenschaftliche Reisen im tropischen Amerika* [Stuttgart, Germany: Cotta, 1870]) did not appear until 1870, in a book that Ratzel said was, unfortunately, little read.

26. Smolka briefly discusses the large monetary value contemporaries assigned to Wagner's natural history collections from his American trip in the later 1850s; the evidence suggests the possibility that he hoped to live off the proceeds when he was unemployed in the early 1860s. See Smolka, *Völkerkunde in München*, 78–81.
27. See, for example, Wagner, "Beiträge zur Naturgeschichte Vorderasiens," in *Reise nach Persien*, vol. 2, app. 2: 282–315; and Wagner, "Beobachtungen über die Fauna der Kaukasusländer und der kolchischen Küste mit besonderer Berücksichtigung der geographischen Verbreitung der Thiere," in *Reise nach Kolchis*, 307–341.
28. Wagner, *Reise nach Kolchis*, 307.
29. Wagner and Scherzer, *Reisen in Nordamerika*, 1:49.
30. Ibid., 1:55; and Wagner, *Der Kaukasus und das Land der Kosaken*, 2:221–222.
31. Wagner, *Der Kaukasus und das Land der Kosaken*, 2:206–213.
32. Ibid., 2:216, 219.
33. Wagner and Scherzer, *Reisen in Nordamerika*, 1:73–74.
34. Ibid., 1:74–77.
35. Ibid., 2:76–77.
36. Wagner, *Die Darwin'sche Theorie*, 63. Wagner briefly discusses the practice of prairie burning in maintaining the prairie plants and keeping out trees in his 1854 book on North America, (*Reisen in Nordamerika*, 3:29–30), but without the explicit territorial interpretation.
37. Wagner, *Reise nach Kolchis*, 91–92.
38. Ibid., 134–135. The village of Elisabeththal no longer exists.
39. Ibid., 88–89.
40. Ibid., 104–107, on p. 106.
41. Ibid., 88.
42. For a different view of pre-1871 German colonies, see Frank Lorenz Müller, "Imperialist Ambitions in Vormärz and Revolutionary Germany: The Agitation for German Settlement Colonies Overseas, 1840–1849," *German History* 17 (1999): 346–368. See also Woodruff D. Smith, "The Ideology of German Colonialism, 1840–1906," *Journal of Modern History* 46 (1974): 641–663.
43. See Wilhelm Roscher, *Kolonien, Kolonial-Politik und Auswanderung*, 2nd ed. (Leipzig, Germany: C. F. Winter, 1856).
44. See Friedrich Ratzel, "Die Gesetze des räumlichen Wachstums der Staaten," *Petermanns Geographische Mitteilungen* 42 (1896): 97–107; and Smith, "Friedrich Ratzel," esp. 67.
45. Andreas W. Daum, *Wissenschaftspopularisierung im 19. Jahrhundert: Bürgerliche Kultur, naturwissenschaftliche Bildung und die deutsche Öffentlichkeit, 1848–1914* (Munich: R. Oldenbourg, 1998), esp. chap. 3.

Three

J. Conor Burns

Negotiating the Agricultural Frontier in Nineteenth-Century Southern Ohio Archaeology

In the early nineteenth century, the Greater Mississippi River watershed was home to innumerable large-scale mound and earthwork constructions generally attributed to an ancient "moundbuilder" civilization. The question of the mounds' origins became central to nineteenth-century theories about the peopling of the New World, and yet by the end of the century almost all these sites had been obliterated by postcolonial development. Construction associated with towns, cities, and transportation routes was a factor, but agricultural practices—especially in the form of plowing—had the most widespread impact. In mere decades, regular cycles of plowing could greatly reduce the largest of mounds, and in the process churn to the surface masses of associated artifacts. Agricultural spaces thus became important gateways to the distant past, where direct confrontations with the archaeological record occurred in a continuous process. In transforming arable land, farmers worked downward into archival landscapes heavily marked by the traces of prior human activity. No wonder that farms became principal work sites for many nineteenth-century American archaeologists, and that farmers became important figures within the developing networks of archaeological field science.

The region of southern Ohio occupied a special place within this context. Due to particularly dense concentrations of elaborate mound and earthwork sites, southern Ohio appeared once to have been the very epicenter of moundbuilder civilization. In the region's broad alluvial bottomlands, especially, the earthen monuments were commonplace topographical features of a landscape eminently suitable for agricultural development. By mid-century, a vast proportion of the total land surface was being farmed, with direct consequences for the ways specimens and other data were collected and archaeological sites accessed, delimited, investigated, and interpreted. The destructive effects of Ohio agriculture brought evidence to light while spurring archaeological efforts to document moundbuilder remains before those traces were irreparably disturbed. By 1880 an amplified sense of crisis over this situation spurred major centralization projects undertaken in the region by both the Peabody Museum of Archaeology and Ethnology and the Smithsonian Institution's Bureau of Ethnology. By around 1890 much more was known about Ohio's ancient inhabitants than had been the case earlier in the century. But this change came at the price of the wholesale loss of numerous in situ archaeological sites.

Many nineteenth-century scholars regarded southern Ohio as one of the most archaeologically significant places in the world, and for that fact alone the region merits the attention of historians. This essay's focus on southern Ohio additionally draws on recent developments in the history of science emphasizing the importance of place and practice in the field sciences.[1] From an analytical point of view, it is a productive strategy to define a geographically limited area and then track the broad range of activities occurring there that affected field practices over time. Recently, Robert Kohler outlined a useful framework for further advancing historical investigations of practices associated with the "collecting sciences," including archaeology.[2] I hope that the present essay might stand at least in part as a contribution to that particular project. Any appreciation of the work undertaken by nineteenth-century archaeologists in southern Ohio—from specimen collecting to excavations—inevitably involves examining what those investigators did on farms as well as how they interacted with farmers. It also necessarily involves acknowledging the fact that agricultural practices to a large extent created the preconditions for collecting in the first place. By looking at these relationships in one region over time, we can gain a better perspective on how collecting was done. My principal goal in this essay is thus to understand the development of archaeological science in relation to the working field spaces of what I will call Ohio's agricultural frontier. I want to elucidate the conditions under which archaeological data was generated, collected, and ultimately interpreted, because those conditions have had long-term repercussions on our current general conceptions of the continent's pre-Columbian past. Indeed, to a large measure, activities that took place on farms shaped the development of archaeology from hobbyist collecting activity to institutionally centralized practice, along with associated transformations in methodology and theory.

On a final introductory note, it must be acknowledged that the account presented in this essay represents but one chapter in a much deeper history. As

William J. Turkel has shown in his exemplary study of the Chilcotin Plateau in British Columbia, land use inevitably brings into conflict a region's materially embedded, multiple histories.[3] The presence of the ancient monuments clearly indicated to many nineteenth century Euro-American settlers in southern Ohio that they were but recent entrants to a built landscape with a long record of human occupation. Very few extant Native Americans lived in the region by this time, and during the rapid agricultural transformation of the state, native voices were not included in archaeological proceedings.[4] Agricultural practices facilitated the emergence of an archaeological science that in turn produced knowledge about the region's longer-term human activities and landscape transformations. In the process, much was subsumed.

The Agricultural Frontier

In a comprehensive mid-century review of the state of American archaeology, Samuel Haven observed that with the "interior of the country tranquilized" following the end of the War of 1812, "a lively spirit of inquiry [had] sprung up in the midst of the antiquities to be investigated."[5] As the librarian since 1838 for the American Antiquarian Society in Worcester, Massachusetts, Haven was not only a fixture of eastern U.S. ethnological circles, he was also in command of one of the largest collections of archaeological literature in the country.[6] For Haven, as for other similarly minded scholars of the day, the major focus of American archaeology was the ancient monuments of the upper Mississippi and Ohio River valleys. He recognized that the growth of the "spirit of inquiry" after 1830 was directly tied to the pronounced increase in the numbers of reported encounters with the archaeological monuments of the Central West.[7]

In southern Ohio, a region explicitly identified at least as early as 1820 by archaeological scholars as the center of an ancient moundbuilding civilization, the end of the war (in 1815) had precipitated significant changes in physical and cultural geography. First of all, the presence of Native Americans within the state was virtually eliminated in all but a few very small reservations in northwestern Ohio.[8] Their physical removal and subsequent absence from areas of the state containing dense concentrations of mounds and earthworks played a significant part in the dissociation of extant American Indians from the area's archaeological heritage. This in turn contributed to the emergence of what recent historians have called a "moundbuilder myth," or the idea that moundbuilders were a distinctly separate and more sophisticated race from Indians.[9] It was also part of a much broader intellectual reconfiguring described by Steven Conn by which Native Americans were not only written out of American history, but were denied possession of a history at all.[10]

Second, changes in federal policies following the end of the war made it easier for Euro-American settlers to lease or purchase land in the state, thus initiating a long period of immigration and economic development unprecedented in U.S. history.[11] As historian Andrew R. L. Cayton recently characterized it, "Ohio recapitulated the history of colonial encounter, conquest, and postcolonial development with breathtaking speed."[12] From the end of the eighteenth century to the middle of the nineteenth, Ohio went from being a territory portrayed as frontier wilderness to being a populous and settled state whose growth rate exceeded that of the entire nation.[13] Euro-American settlement began in the southern part of the state, along the Ohio River, and proceeded northward. By 1860 the population was over 2.3 million, third behind only New York and Pennsylvania.[14] Cincinnati grew from a population of under 10,000 in 1820 to 161,044 in 1860, and was considered to be one of the most ethnically diverse places in the world.[15] Early intellectual leaders such as Cincinnati doctor Daniel Drake saw Ohio as a utopian social experiment in which immigrant groups of unprecedented religious and ethnic variety transformed the land into an economically productive paradise.[16] Drake was an archaeological enthusiast who founded the Western Museum Society in 1818. Short-lived as it was, this organization's museum was the first repository for archaeological materials in the West.[17]

Historians have shown how the range of specific agricultural practices in nineteenth-century Ohio reflected the diversity of the state's immigrant groups, unified mainly by their shared experience of hard labor.[18] Significantly, Ohio farmers as a whole proved remarkably resistant to mechanization in the form of technologies such as the threshing machine.[19] This seems to have been especially true of those who worked in the more deeply furrowed landscapes of the southern half of the state. Ohio's agriculturally derived economic output after the Civil War was thus not tied to the emergence of large-scale, mechanized farming, as it would be in prairie states further west, but came through sheer dint of numbers: Ohio was a state brimming with multitudes of small, productive farms. According to historian George Knepper, by 1880 there were 247,189 farms in Ohio. The average size was ninety-nine acres, and an overwhelming majority of them were worked by their owners and not by tenants.[20] What tied all this activity together and allowed the agricultural economy to burgeon were extensive transportation networks of roads, canals, and railroads. On nineteenth-century maps of Ohio, many of the state's cities and towns appear as the hubs of giant wagon wheels, with various transportation routes radiating outward in all directions. Ohio as a whole became a principal transportation hub linking East and West.

The relationship in the early American republic between agricultural commercialization and transportation improvements, on the one hand, and the formulation of Indian policy (along with the emergence of attitudes toward native peoples), on the other, is a complex one, and only recently have historians begun to explore it in depth and expose explicit links.[21] In Ginette Aley's fine-grained frontier thesis, for instance, the Jeffersonian vision of a national agricultural economy directly informed land acquisition and development policy in the West for much of the nineteenth century.[22] Aley's frontier is a borderless zone where rapidly expanding transportation networks fueled the growth of that economy and simultaneously facilitated the removal of Indians through the very land that had been taken from them. "There is still much to be learned," she writes, "about the process by which Euro-Americans assumed hegemony over native peoples in other areas beside the battlefield."[23] In the tranquilized interior of southern Ohio, to borrow Samuel Haven's language, hegemony had indeed been achieved, and the plow was at least partly responsible for the fact that Native Americans no longer lived there. By the time Haven's words were printed in 1856, however, it had also become patently clear that the piecemeal activity of plowing was erasing the archaeological traces of Ohio's ancient inhabitants.

Even in adopting a refined and much revised understanding of the American frontier such as that offered by Aley, we tend to think of the frontier in horizontal, geographic terms. As western lands were colonized, Native Americans were displaced further west. But in the archaeologically rich and agriculturally fertile Ohio valley, another frontier developed after the region was settled by Euro-Americans. This was a vertical boundary between present and past, between history and "prehistory," between what was known or assumed about Native Americans and the archaeological record pertaining to them. Agricultural engagement with the land leveled mounds and dislocated vast amounts of archaeological material, and Ohio's farms became focal points for archaeologists of every stripe, from hobbyist collectors to scholarly investigators.

Agriculture and Archaeology, Farmers and Archaeologists

Awareness of the agriculturally based destruction of southern Ohio's archaeological monuments began in the early decades of the century. In the first substantial scholarly account of them, Caleb Atwater, a corresponding member of the American Antiquarian Society, sounded an alarm about the damage being done.[24] Atwater was also the postmaster of Circleville, Ohio, a town whose very name derived from the ancient earthworks it was built atop. His "Description of the Antiquities Discovered in the State of Ohio and Other Western States" was published in the first (1820) volume of *Archaeologia Americana*, the transactions of the American Antiquarian Society.[25] It represented the most extensive catalog to date of works and synthesis of work done on them. It also set the tone for the treatment of moundbuilders for the remainder of the century, as Atwater clearly demarcated this class of antiquities from the "Antiquities of Indians of the present race."[26] In a lengthy analytical section of his study, Atwater made comparisons between select moundbuilder artifacts and Old World symbolisms to make a case for the Asiatic origin of the moundbuilders, meaning that the pieces did not apply to what were in his view the more degraded Indians of recent history.[27] Indians, that is to say, were not in any way descended from the moundbuilders.

Within two decades, the Cleveland topographic engineer Charles Whittlesey had been charged with surveying Ohio's ancient works as part of the Geological Survey of the State of Ohio.[28] Just the fact that this task had been included as part of the survey of the state's geology attests to Whittlesey's claim that, literally, "no portion of Ohio appears to be destitute of ancient tumuli and embankments."[29] Ancient Native Americans had altered and shaped the landscape to such a significant degree as to leave Euro-American surveyors thinking in large-scale geomorphological terms rather than in terms of individual archaeological sites. (Whittlesey, needless to say, did not refer to the remains of ancient Native Americans but to "ruins of a lost race.") Two other striking features issued from the topographic survey section of the report—one, a denunciation of Atwater's survey for being inaccurate, and two, an urgent warning about the pace of archaeological destruction. After acknowledging how few of the works had been surveyed or described in any manner whatsoever, Whittlesey wrote that as a whole "their forms and dimensions are fast disappearing under the operation of the plough and the spade."[30] He continued, "For it is in the rich valleys of the Miami, the Scioto, and the Muskingum, where the modern agriculturist now cultivates the soil, that an ancient people, more numerous than the present [Euro-American] occupants, pursued the same peaceful avocation, at least ten centuries ago; and upon the sites of modern towns within these valleys . . . the ancients located *their* cities, of which distinct traces exist."[31]

As the "learned abroad" (read: armchair scholars from outside Ohio) increasingly paid attention to Ohio's mounds, it became that much more essential, warned Whittlesey, to get the facts on the ground right before they were plowed out of existence. The views of prominent Massachusetts geologist Edward Hitchcock offer an extreme example of the kind of misperceptions causing worry. Through the late 1830s, Hitchcock—who by one account had never seen an Ohio mound in person—claimed that Ohio's mounds and earthworks were simply too numerous and too large to have all been man-made productions. Instead, he argued that natural processes of flooding and sedimentation were the likely origin of a majority of the works.[32]

One of Whittlesey's field collaborators in the 1830s and 1840s was the southern Ohio doctor Edwin Davis.[33] Davis was a keen archaeological enthusiast who shared Whittlesey's concern over the destruction of mounds. His formative archaeological experience came in the mid-1830s as a result of turnpike construction in the vicinity of Bainbridge, his home at the time. Davis studied many of the mounds bisected by the roadwork, with the aim of understanding how they had originally been built. This work drew the attention of Hitchcock, who refused to accept Davis's determination as to the artificial nature of nonfunerary mounds (and this despite overwhelming evidence to the contrary).[34] Without direct experience, Hitchcock simply could not believe that human beings had made such a large-scale impact on Ohio's topography.

As Whittlesey's survey work got underway, East Coast ethnologists had not only become interested in the ancient works of the Ohio and Mississippi River valleys, they had made the riddle of the moundbuilders into a central problematic of American ethnology.[35] Two figureheads within this consolidating community who were especially interested in the mounds were the monogenist philologist Albert Gallatin and the polygenist anatomist Samuel Morton. Gallatin wanted corroborating archaeological evidence to bear out his theory of the Asian origins of New World aboriginal languages, while Morton wanted to acquire moundbuilder skulls supporting his theory of a truly autochthonous American race.[36] Thus, within the theoretical frameworks developed by Gallatin and Morton, mound archaeology acquired a degree of scientific prestige it had not had before. At the same time, the remaining intact archaeological evidence came increasingly under threat of obliteration.

The next major document of southern Ohio mounds to make that threat apparent was *Ancient Monuments of the Mississippi Valley*, a monograph coproduced by Ephraim Squier and Whittlesey's friend Edwin Davis. Squier was a New York writer who had relocated in the mid-1840s to Chillicothe, Ohio, a town to which Davis had also moved. The two men began collaborating on a project they envisioned as a large-scale archaeological survey unlike anything previously completed. They drew the attention of both Gallatin and Morton, and Gallatin's American Ethnological Society played a role in funneling the project to Joseph Henry, secretary of the fledgling Smithsonian Institution. Henry not only agreed to publish the results, he made *Ancient Monuments* the flagship inaugural volume of the Smithsonian Contributions to Knowledge series. Henry, a physical scientist, wanted this to serve as a model for future Smithsonian science.[37]

Ancient Monuments of the Mississippi occupies a central place in the history of nineteenth-century American archaeology. The Squier and Davis collaboration has been addressed in detail by a number of historians, although the full impact of the work—in terms of how it codified particular standards of archaeological representation and formalized a certain classificatory language for mounds—remains to be examined.[38] The two men performed many field surveys themselves, but they also drew extensively on the survey work of a large number of associates, including Charles Whittlesey. One thus gets a sense of the breadth of the networking involved—no two people alone could possibly have completed such a large project. The 342-page, quarto-sized published volume contained 106 full-page plates of plan-view maps of ancient works, as well as 207 woodcut engravings depicting profile sections of mounds, artifacts, and skeletal specimens. In purely visual terms, nothing like this had ever been published before in archaeology. And because it bore the imprint of the Smithsonian, it became the accepted authoritative account of the mounds for several decades and was widely hailed in the United States and abroad as a landmark of archaeological scholarship.[39]

One of the most misleading aspects of *Ancient Monuments* concerns the plan-view depictions of ancient works. In these maps, the traces of modern development—towns, roads, farm buildings—fade into the background against the bold, hyper-delineated earthworks and mounds (see figures 3.1 and 3.2). Even nature itself—rivers, streams, and forests—recedes into a ghostly backdrop on which the ancient works appear permanently embossed. In reality, of course, just the opposite was the case, and if anything in these maps was in the process of being reduced to an incorporeal existence, it was the archaeological remains. Many of the sites documented in *Ancient Monuments* did not survive the subsequent three decades, let alone the century, and Squier and Davis themselves were acutely aware of the rate of destruction.

In the prefatory portion of the monograph, the authors called attention to the extent of the problem, and as comprehensive in scope as their own work was, they also here acknowledged the utter insufficiency of their effort in the face of the full task. "The importance of a complete and speedy examination of the whole field," they wrote, "cannot be over-estimated." Further, "the operations of the elements, the shifting channels of the streams, the levelling hand of public improvement, and most efficient of all, the slow but constant encroachments of agriculture, are fast destroying these monuments of ancient labor, breaking in upon their symmetry and obliterating their outlines. Thousands have already disappeared, or retain but slight and doubtful traces of their former proportions."[40]

Squier and Davis's astonishing plan view of the area around Chillicothe (see figure 3.3) captures at once many of these concerns. The Scioto River meanders southward through soft alluvial soil. Numerous lower-order waterways drain into it. The town itself hugs a sharp bend in the river, and a number of roads emanate from it in every direction. The Ohio and Erie Canal parallels the

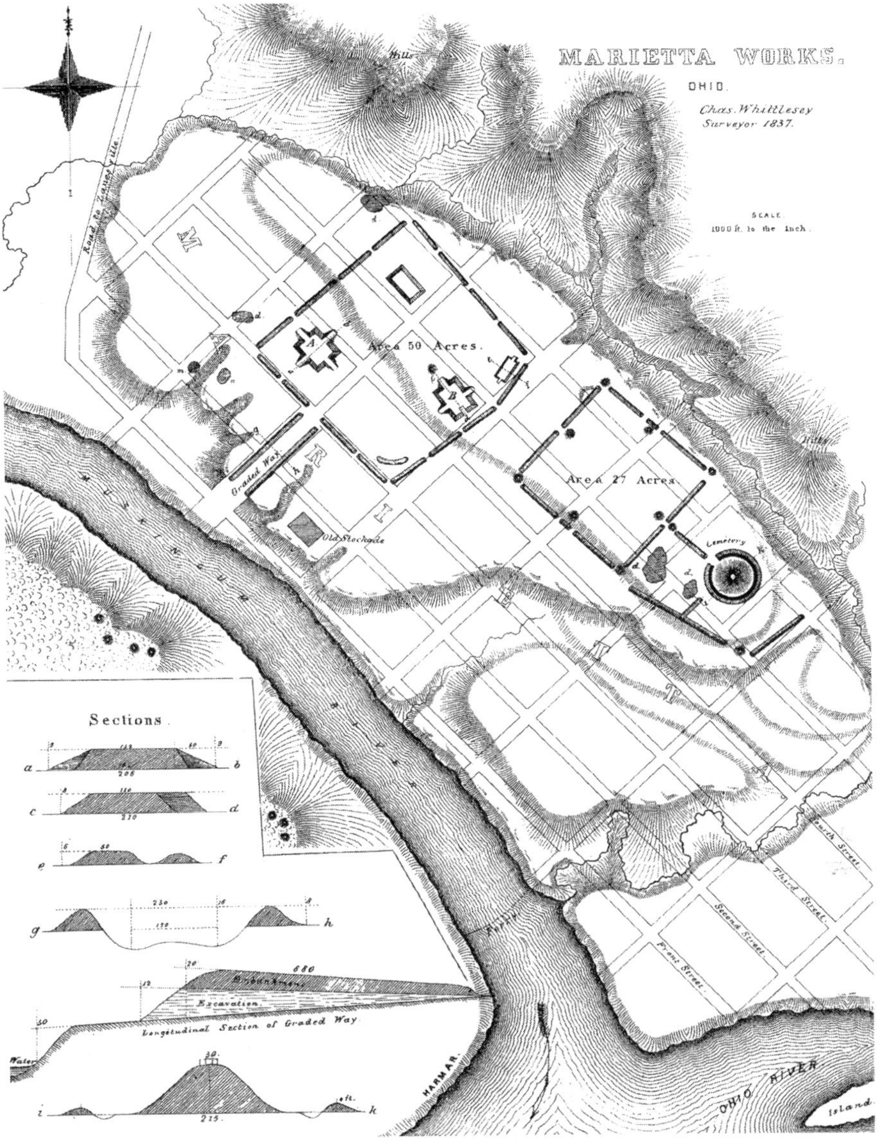

Figure 3.1

The "Marietta Works." The earthworks appear as though superimposed onto the town's ghostly streetscape, which has been depicted without any of Marietta's modern structures. Not to original scale.

From Ephraim G. Squier and Edwin H. Davis, Ancient Monuments of the Mississippi Valley *(1848; repr., Washington, D.C.: Smithsonian Institution Press, 1998), plate between pp. 72–73.*

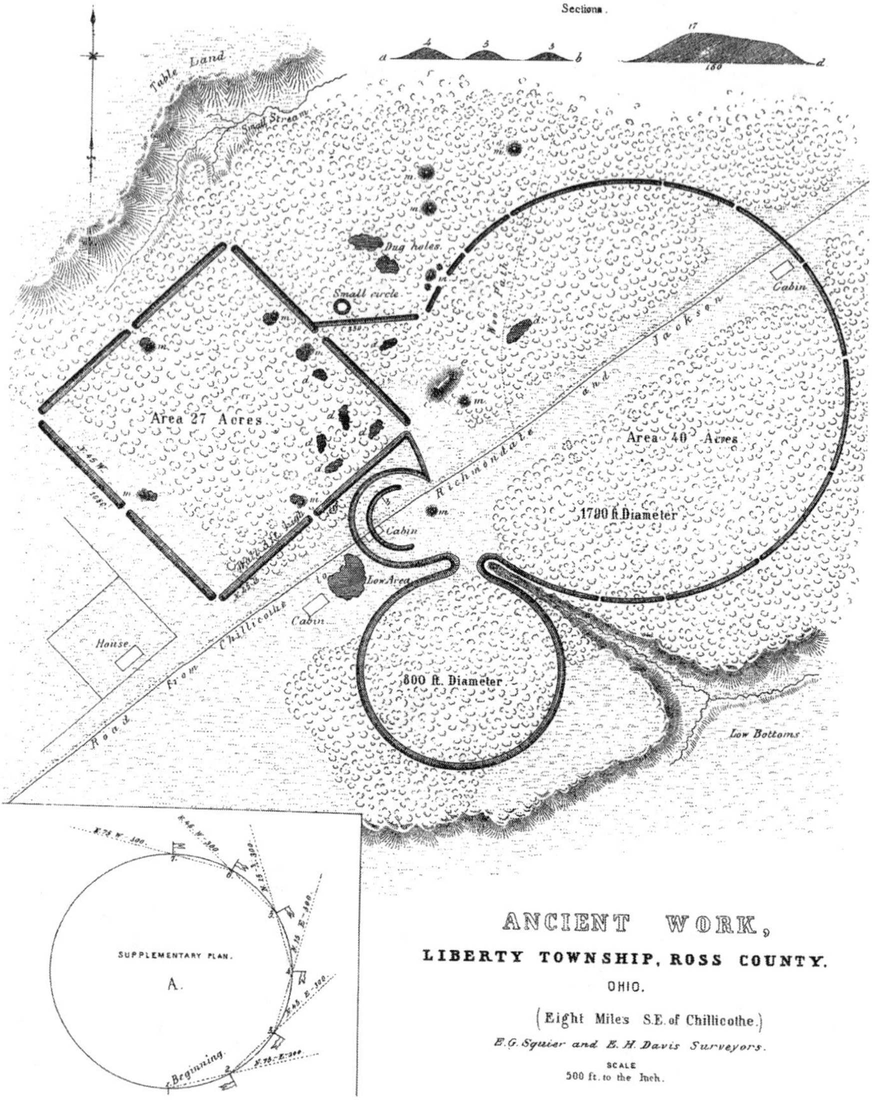

Figure 3.2

The "Liberty Works," an earthwork and mound complex that demonstrated the moundbuilders' supposed advanced knowledge of geometry. Not to original scale.
From Squier and Davis, Ancient Monuments of the Mississippi Valley, *plate between pp. 56–57.*

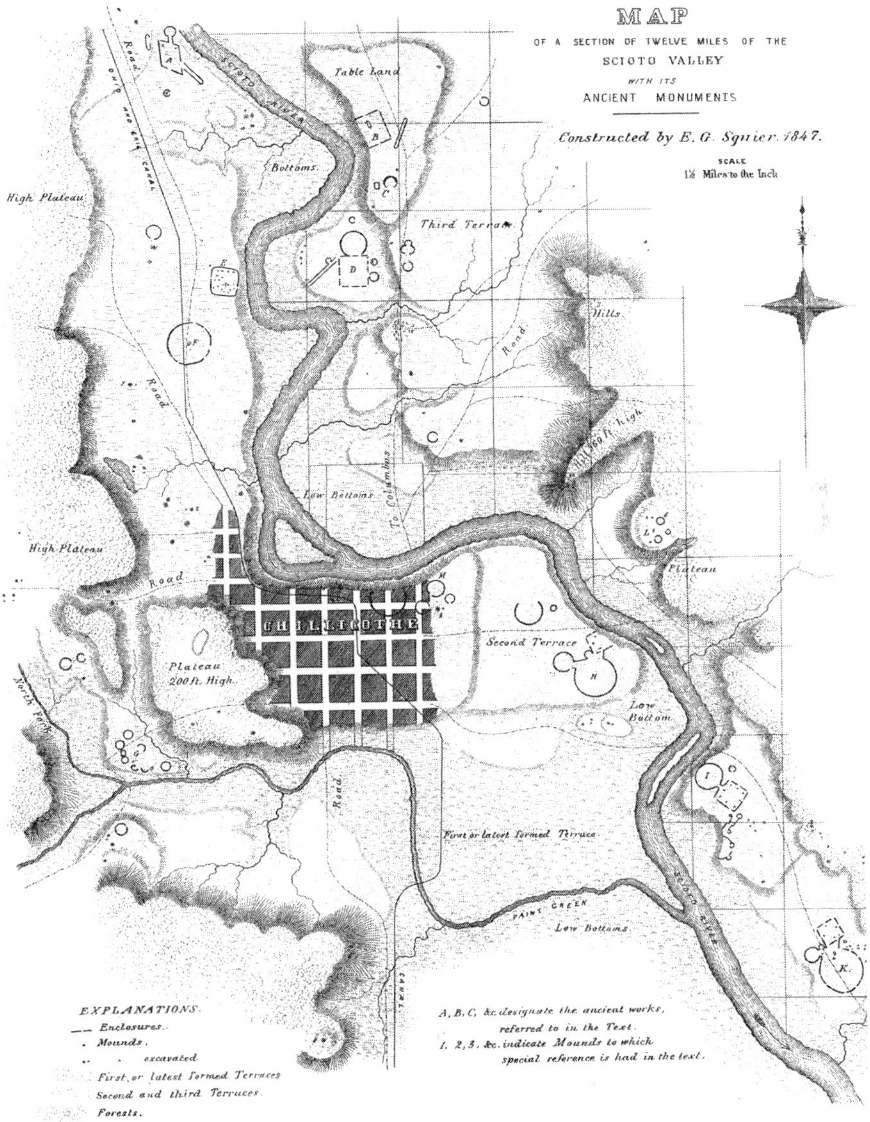

Figure 3.3

Ancient works in the vicinity of Chillicothe. Not to original scale.

From Squier and Davis, Ancient Monuments of the Mississippi Valley, *plate between pp. 2–3.*

river on the west. And everywhere one looks, the landscape is replete with the traces of moundbuilder civilization. Many works have roads or the canal running through them. On associated detail maps, the authors noted the "fine, arable soil" of the lower-river terraces. Indeed, in the broad Scioto Valley lay some of the most fertile, cultivable soil to be found anywhere in the state.[41]

The appearance of *Ancient Monuments of the Mississippi Valley* marked the emergence of a distinct pattern in nineteenth-century American archaeological discussions. In the post–Squier and Davis world of American archaeology, one finds a standard refrain repeated in the literature—the ancient works are of paramount archaeological significance, they are being plowed out of existence, and those remaining must be properly documented before it is too late. But how to go about such an overwhelming task? Who would do it? Who could? Squier and Davis thought it "too great an undertaking for private enterprise to attempt," further stating that it "must be left to local explorers, to learned associations, or to the Government."[42]

And so, for a time after mid-century, the quasi-governmental Smithsonian Institution seemed to offer the best hope for coordinating efforts at least to document the mounds of the American Midwest. In addition to publishing a number of subsequent archaeological monographs in the Contributions to Knowledge series (including two by Charles Whittlesey), the more regularly published *Annual Reports* of the Smithsonian proved to be a useful means of disseminating and eliciting information.[43] Ethnology and the mounds took up a striking proportion of space in the lengthy introductory secretary's reports in nearly every volume of the *Annual Reports* throughout the 1850s to 1870s. In these, Joseph Henry frequently called attention to the deleterious effects of plowing. He also used the platform for nurturing a body of field correspondents who could utilize and further distribute circulars containing instructions on how to examine and document the ancient monuments.[44] The secretary asked that brief individual accounts be sent on to the Smithsonian, where they might be published within the Ethnology section proper of the *Reports*. Henry envisioned a system by which this information would be processed by some requisite expert and added to the framework established by Squier and Davis. In this manner, he thought, a complete record could be made rather efficiently, and it would then become part of a much bigger systematization and classification of North American archaeology and ethnology in general.

In the long term, however, Henry's plan was bound to fall short of what was needed. The availability of economically productive farmland continued to lure immigrants to Ohio, and, if anything, the rate of agricultural destruction continued to rise unabated simply by virtue of the fact that more land was being farmed. The Smithsonian's piecemeal data collection efforts could not possibly keep up, and indeed many of the incoming accounts described works already much damaged or destroyed. It should be pointed out that Henry's plan was strictly a documentation strategy. It did not any way include provisions for the protection of archaeological sites other than to promote some level of awareness in print, nor was Henry himself very interested in any kind of large-scale museum building. The systematic collection of specimens associated with the mounds was simply not within the Smithsonian's purview at this point in time, and the disruption of the Civil War followed by a devastating fire presented serious setbacks to institutional operations.[45]

Nonetheless, through the 1860s and 1870s, the deleterious impacts of agriculture remained a problem of central prominence for many archaeologists. So, too, did the theoretical questions associated with the moundbuilders, which by this time had been tempered by the emergence of post-Darwinian evolutionist anthropology and by rampant debates on both the antiquity and unity of humankind in both Europe and the United States.[46] As archaeology in the United States became increasingly infused with an explicitly scientistic attitude—along with an accompanying rhetoric of archaeology as science—farmers began to take on a much more visible presence within archaeological practices and discourses. Earlier, farmers had been largely anonymous participants and were rarely if ever acknowledged in published accounts. Squier and Davis, for instance, needed to associate with farmers to have accessed many sites. Field transactions of this sort, however, were simply not documented in texts such as *Ancient Monuments*, nor generally were they within the detached notices of fieldwork reported in the Smithsonian *Annual Reports* for some time through the third quarter of the century. By about 1880, this would change decisively as farmers themselves—and not only the effects of their practices—entered much more explicitly into archaeological proceedings.

Increasing Institutional Presence on the Agricultural Frontier

One main reason for this shift had to do with the emergence of two particular institutions capable of undertaking large-scale, coordinated archaeological field projects in the Ohio Valley. One was the Peabody Museum of American Archaeology and Ethnology in Cambridge, Massachusetts, founded in 1866 thanks to a large donation to Harvard.[47] It was the first American institution dedicated exclusively to archaeology and ethnology. After a fitful first decade, the mantle of director settled onto Frederic Ward Putnam, a former student of zoologist Louis Agassiz. Under Putnam's guidance, the Peabody quickly acquired a reputation as one of the world's leading archaeological museums. By the end of the 1870s, Putnam was convinced of the importance of Ohio valley archaeology, and shortly thereafter he was overseeing fieldwork in the region. For much of the 1880s, he became increasingly involved in southern Ohio archaeology until his work there ended as he became involved in preparing the anthropological exhibits for the Chicago World's Columbian Exposition. The other was the U.S. Bureau of Ethnology, formed in 1879 under the same congressional bill that established the U.S. Geological Survey and placed John Wesley Powell in charge of both.[48] The Bureau operated essentially as an adjunct to the Smithsonian, where its headquarters were located, and where the Smithsonian's new National Museum of Natural History provided storage, display, and research space for collections. The presence of the Bureau thus added much-needed focus, direction, and oversight to the Smithsonian's ethnological efforts. Powell, famous for his Department of the Interior western survey work, was more interested in linguistic and ethnographic projects, but pressure mounted to confront the problem of the mounds. He created a Mound Exploring Division that was fully operational by 1882 under the supervision of Cyrus Thomas, another western survey veteran.[49] Thomas was more or less expected to complete a full survey of the Greater Mississippi valley mounds in less than a decade.

The pronounced involvement of both the Peabody Museum and the Bureau of Ethnology had a considerable impact on what went on in the field. While the overall work done through these institutions embodied strikingly divergent investigative styles, a shared aim of methodical centralization in the face of the widespread loss of archaeological data drove the projects.[50] To this end, Putnam and Thomas promoted archaeology as a science and proactively engaged local archaeological practitioners—individually as well as through archaeological and historical societies—as a means of building networks of access to archaeological sites and materials. A number of such individuals became correspondent field-workers, some of whom were paid expenses or even a salary for archaeological work done in the name of the Peabody or the Bureau.[51] Institutional figureheads also became more willing to purchase collections, sometimes at substantial expense, exchanges that did not go unnoticed among local collectors. Despite these shared aims and a veneer of gentlemanly cooperation and courtesy, however, considerable inter-institutional competition emerged as Putnam and Thomas each sought to access sites and specimens.[52] On the ground in southern Ohio, the net effect of these Peabody and Bureau initiatives created an intensified and often uneasy dynamic for practitioners.[53]

Within this particular context, the relations between archaeology and agriculture, between archaeologists and farmers, became significantly more interconnected in the 1880s than they had been earlier. First and foremost, archaeologists of the Peabody Museum and the Bureau of Ethnology recognized that the agricultural destruction of mounds and earthworks had reached a devastating crescendo. This became a common trope within the published archaeological literature and within other forms of archaeological discourse associated with the Peabody and Bureau work of the decade. But it was not mere hyperbole: to find a mound or earthwork that had not been seriously damaged by the plow became a rare exception to the rule.[54] And yet next to nothing was being done—nor could be done—as hundreds of thousands of small-time farmers in Ohio toiled away at their modest plots of land, each successive season of cultivation and animal grazing steadily wearing away at the archaeological remains. In some particularly archaeologically rich areas, the statistics of agricultural land use had attained seriously impressive numbers. For example, in Ross County, of which Chillicothe was the seat, over 50 percent of the total land surface was under cultivation or in pasture by 1887, according to one source.[55]

Putnam was especially proactive—and provocative—in broadcasting his concerns. Within Peabody publications and in correspondence with members of the newly organized Ohio State Archaeological and Historical Society (OSAHS) as well as various Ohio newspapers, Putnam regularly and relentlessly emphasized the problem.[56] He frequently appealed to Ohioans' own sense of pride to save the monuments. To him this seemed the only viable solution. In the first volume of the widely read journal of the OSAHS, for instance, Putnam expressed in no uncertain terms what was at stake: the widespread loss of an archaeological heritage as important to the United States as the pyramids were to Egypt, all as a result of careless development.[57] In clearly identifying agriculture as the prime culprit, he wrote, "Shall such vandalism, such shame, be laid to Americans of this century?"[58] People of Ohio, do something to preserve at least some of the works, Putnam urged, and do it now. The Peabody director set a truly exceptional example in what can best be described as a campaign to raise awareness about the loss of Ohio's mounds, culminating in the Peabody's purchase of Serpent Mound in 1887.[59]

During the 1880s, Squier and Davis's *Ancient Monuments of the Mississippi Valley* became for many field-workers a benchmark by which to gauge the agricultural damage inflicted on the mounds over the previous thirty to forty years. Individuals in the field often struggled to find any trace of sites documented in *Ancient Monuments*. In examining works on the Edwin Harness farm in the Scioto Valley and during a reconnaissance survey of the Portsmouth Works at the mouth of the Scioto River, for instance, Putnam and his field collaborator Charles Metz found that significant portions of these sites (originally described by Squier and Davis) had been heavily damaged if not completely obliterated by plowing.[60] James Middleton, one of the paid field-workers for Thomas's Mound Exploring Division, repeatedly encountered similar circumstances in his work for the Bureau of Ethnology.[61] Middleton was one of a number of Bureau field archaeologists who had been systematically re-surveying works first surveyed by Squier and Davis. The veneer of *Ancient Monuments* as the standard and most reliable account of the Ohio mounds had begun to wear thin as field-workers in the 1880s noticed that Squier and Davis had inaccurately emphasized the geometrical regularity of many earthworks.[62] The moundbuilders' supposed knowledge of geometry had been used to support the theory that they were an advanced race with no lineal relationship to living American Indian tribes, a theory that Powell and Thomas explicitly sought to discredit. And now, as Middleton and other Bureau workers were finding, many of the works described by Squier and Davis had been rendered so indistinct by plowing as to make re-surveying them in any accurate way very difficult.

At the same time that the detrimental effects of farming practices had reached worrisome levels for archaeologists associated with the Peabody and the Smithsonian, farmers themselves became extremely valuable assets for archaeologists, especially for those associated with the Peabody or Bureau of Ethnology work. Building and maintaining good relations with farmers gave archaeologists access to sites and specimens not otherwise possible. Putnam's success in establishing a strong Peabody field presence in southern Ohio, first in the Little Miami River Valley near Cincinnati then in a number of other important locations in the region, came largely as a result of such efforts. In fairness, Charles Metz deserves considerable credit for this.[63] Metz, a Cincinnati doctor heavily involved in the local archaeological scene, had already been overseeing investigations of sites in the Little Miami Valley when Putnam began collaborating with him in the early 1880s.[64] The Little Miami sites were all on farms with whose owners Metz and Putnam had established a good rapport. In his *Peabody Reports*, Putnam could openly state that numerous farm owners in the valley had granted "exclusive right of exploration" to the Peabody Museum.[65] Metz consciously went out of his way to be courteous to farmers in order to guarantee ongoing archaeological fieldwork at particular locations.[66] Putnam's cordial relationship with one farming family ultimately allowed him to purchase Serpent Mound and perform three successive field seasons' worth of work on this most enigmatic of mounds.[67] Putnam utilized sources such as the *Chillicothe Leader* newspaper to establish relations with farmers specifically to access sites.[68] The sort of intensive, long-term field investigations typically conducted by Metz and Putnam—work usually done during warm months—was at least somewhat disruptive to farmers, so it is significant indeed that many farmers were willing to aid them in such ways. As a result, Putnam and Metz laid claim to many of southern Ohio's most important mound sites. No wonder that Smithsonian and Bureau of Ethnology authorities became seriously concerned about Putnam "monopolizing" Ohio archaeology by mid-decade.[69] Thomas subsequently directed more of the Mound Exploring Division's resources and personnel to Ohio.

Farmers were archaeological assets in other ways in the 1880s. Charles Smith was a part-time teacher in the Columbus area and an archaeological enthusiast who desperately wanted to work for the Bureau of Ethnology. He got a chance largely on the basis of his claimed associations with many farmers.[70] Smith's archaeological track record was completely unproved—no publications of any sort, no involvement with established local organizations or networks. On the basis of experience alone, he had little hope of being hired as a Bureau of Ethnology archaeologist. But Smith perceived very clearly the value of his agricultural associations. In letters to Thomas, Powell, and Smithsonian Secretary Spencer Baird, Smith wrote of being able to get more work done for the money than anyone else because he both knew the terrain and was friends with many farmers who would help him out. This was exactly what Thomas needed to establish a better Bureau presence in Ohio, and so Smith was brought on board. For Smith, then, farmers literally served as a form of capital guaranteeing his entrance into the world of paid institutional archaeology. And once in, many of those farmers granted invaluable access to archaeological resources on their lands. Smith's first major job for the Bureau—a survey of the Flint Ridge quarries in central Ohio—required such access.[71] This project resulted in a respectable publication that appeared in the *Smithsonian Annual Report* for 1884. The article helped Smith establish his reputation as an archaeologist, and he worked as one of the Bureau's principal field agents for several years.

Aside from being gatekeepers to archaeological sites, farmers also became important sources for specimens. This situation, too, was not new by the 1880s—many objects acquired or described by Squier and Davis, for instance, came via farmers—but within the context of what historian Mark Barrow has identified as the post–Civil War emergence of entrepreneurial natural history, farmers took on much more privileged roles as specimen sources.[72] "Entrepreneurial archaeology" in the Ohio valley peaked in the 1880s, largely as a by-product of the Peabody and Smithsonian Bureau of Ethnology projects.[73] Since centralization for both institutions at this point in time explicitly included building museum collections, desirable specimens became much more valuable than they had been earlier in the century. To illustrate, in the 1860s no American institution—including the Smithsonian—was willing to purchase even the world-famous Squier and Davis cabinet. Davis ended up selling the collection to English entrepreneur William Blackmore, who featured the objects in his new museum in Salisbury.[74] In the 1880s, enough had changed so that even Ohio farmers regularly profited from the sale of artifacts, very often to individuals associated in some way with the Peabody or Smithsonian.

Within the context of a growing market for moundbuilder artifacts, then, one glimpses the extent to which farmers held in their collective possessions a vast storehouse of archaeological materials. For example, Charles Smith, mentioned above, reported to Thomas that most farmhouses in his area of Ohio contained collections of specimens "gathered from the soil" that farmers were more than willing to sell. "I have been accustomed to take things, if worth the money asked," he wrote, "and adding to my own collection or selling them. I will send them to you at what they cost, if you want to accumulate in that way."[75] Smith often kept the choicest items for his own collection, selling duplicates to the Smithsonian. Warren Moorehead, another ambitious, young archaeological enthusiast from central Ohio looking to find a place among the ranks of paid institutional workers, had a similar approach to collecting from farmers. In correspondence with Spencer Baird, Moorehead provided a list of various specimens for sale by farmers, and he offered his services as purchaser for the Smithsonian Institution in the region. Moorehead described a number of representative individual items for sale for five or ten cents each, as well as a collection he had obtained worth thirty to forty dollars that he could part with for ten or twelve.[76] The case of W.C. Hampton offers a slightly different perspective. Hampton operated a small Ohio agricultural supply business that seems to have doubled as an artifact dealership. His flyers not only advertised specimens for sale, but also offered to exchange goods such as seeds for "moundbuilder relics."[77] Hampton saw himself as an intermediary between the agricultural sources of artifacts and the Smithsonian, writing directly to Powell in hopes of selling items to the Institution.

Conclusion

Nowhere do the complex associations between agriculture and archaeology come through more tellingly than in the case of Serpent Mound. Situated atop a prominent bluff in a remote, south-central part of the state, this quarter-mile-long serpent effigy earthwork and associated complex of burial mounds had been protected by a covering of forest in the early part of the century.[78] In the 1820s, the site was located on farmland owned by William Hamilton.[79] Squier and Davis first documented it in *Ancient Monuments of the Mississippi Valley*, after which the "Great Serpent" of Adams County became one of the most famous archaeological sites in North America. Ethnologists vigorously disagreed over what the work was meant to represent—some thought it a giant serpent in the act of devouring an egg—but they uniformly interpreted it as a religious site.[80] Many thought that this particular place was where the moundbuilders gathered to perform their most sacrosanct rituals. Interpretations tended to fix especially on the serpent's topographic harmony with the resident landform—seemingly clear evidence of a kind of animistic impulse on the part of the site's architects. Serpent Mound came to be regarded as a crucial piece of evidence for locating the precise developmental attainments of moundbuilder religion in comparison to other known religious systems within which the serpent symbol made an appearance.

When Frederic Ward Putnam visited the Serpent in 1883, he found that it had been much damaged since the 1840s.[81] No longer protected by overgrowth, the entire site had been cleared and was under active cultivation.[82] Putnam saw that the plow, especially, had been responsible for much of the damage to the monument. In late 1886, when Putnam next returned to the site, he found that the damage had accelerated greatly, and he immediately began negotiations with current landowner and farmer, John Lovett, to purchase and preserve the site in the name of the Peabody Museum. From the time of his first visit, Putnam cultivated a careful relationship with the Lovett family, which served him well throughout the property negotiations and subsequent field investigations. Lovett himself possessed only a lukewarm interest in archaeological matters, but he became committed to aiding Putnam. Within a year, the transaction was complete, and Putnam began the first of several field seasons' worth of explorations on the site. By the turn of the decade, Serpent Mound Park was opened to the public, with funds for both its purchase and upkeep generated via a subscription campaign that had been organized in Boston.[83]

In the end, the Peabody director was able to conduct what stands as one of the most intensive and methodical investigations of nineteenth-century American archaeology. His excavations, generally conducted with the aid of Charles Metz and a number of others, revealed a vast, multicomponent archaeological site. Putnam examined the manner of construction of the serpent effigy, and he explored in detail a nearby associated burial mound. He also located a vast subsurface village component to the site, as well as a number of other scattered burials. He found a striking diversity of artifact types, burial styles, and human osteological specimens, indicating to him that numerous different peoples had used the site over a vast span of time. In other words, it was not the product of a single group of "moundbuilders," but rather the result of a much more complicated set of human migrations in the prehistoric past.[84] This conclusion was significant.

Putnam was no "separate race" theorist when it came to moundbuilder identity, but his problematization of North American prehistory stood in stark contrast to the views being promoted by the Bureau of Ethnology by the end of the

1880s. For John Wesley Powell and Cyrus Thomas, moundbuilders were direct, lineal ancestors of particular Native American tribes.[85] In taking this stance, however, the Bureau archaeologists denied any great antiquity to the natives of "prehistory," and they thought that approaches such as Putnam's unnecessarily complicated inferential frameworks for making claims about the past. Putnam, on the other hand, thought that Bureau archaeology oversimplified the situation and failed to account for the immense diversity and complexity that characterized North American prehistory. The differences in outlook could not have been more pronounced.

In southern Ohio, farms were places where some of the most important questions of nineteenth-century archaeology were framed and addressed. I have characterized working farm spaces collectively as an agricultural frontier, and argued throughout this essay that activities here were deeply determinative to the proceedings and results of archaeological practice. At the most obvious level, archaeological practices—from the activities of local collectors to the Peabody and Bureau of Ethnology projects—emerged in response to the increasingly deleterious effects of agriculture on the mounds. Agriculture was by no means the sole destroying force, but it was the one with the broadest impact and it attracted the greatest attention from many archaeologists.[86] However else one might wish to characterize the work orchestrated by either Putnam or Thomas, ultimately these were salvage efforts to document the rapidly vanishing traces of North America's archaeological past.[87] But the same might have been said for work done four decades earlier by Squier and Davis, and by others along the way as well. American archaeology emerged predominantly as a salvage science, meaning that its data and its objects of study were determined by the destructive course of postcolonial development.

The associations between agriculture and archaeology became considerably more nuanced, however, in the various activities that took place on individual farms. Contingencies at the local level could have great significance for shaping investigators' theoretical views about the character of prehistoric North America, as in the case of Putnam's work at Serpent Mound. In general, building and maintaining good relations with farmers was essential for archaeologists associated with the Peabody or the Bureau of Ethnology, and those who did it well were rewarded. In some cases, as I have shown, associations with farmers proved to be more valuable forms of archaeological experience than those we might typically associate with fieldwork. The fact that farmers became more visible participants in institutional archaeology in the 1880s somewhat challenges our conventional notions about the emergence of expertise and authority in science. The networking required by Putnam and Thomas could not have functioned without the input of farmers. In fact, it worked more efficiently and successfully when farmers' efforts were acknowledged rather than excluded.

Very few of Ohio's ancient monuments survived the nineteenth century. Only Serpent Mound and a handful of others have been designated as state or national historic sites. Today, archaeologists attribute the Ohio mounds variously to the Adena, Hopewell, Fort Ancient, or Mississippian cultures, and they believe the works were constructed over approximately two and a half thousand years, ending in the sixteenth century. What today's archaeologists know about these various native peoples derives largely from evidence produced in the nineteenth century under conditions such as those described in this essay. In seeking a critical understanding of those conditions—that is, of the complex associations between archaeological and agricultural practices—we might better appreciate the means by which archaeologists in turn have come to construct narratives about the past.

Notes

I would like to thank all the organizers and participants involved in the "Knowing Global Environments" conference, especially Jeremy Vetter and Robert E. Kohler. I also thank Katharine Anderson and Alison Wylie for detailed feedback on ideas presented in this essay.

1. See, for instance, Henrika Kuklick and Robert E. Kohler, eds., *Science in the Field, Osiris* 11 (1996); Samuel J. M. M. Alberti, "Amateurs and Professionals in One County: Biology and Natural History in Late Victorian Yorkshire," *Journal of the History of Biology* 34 (2001): 115–147; and Robert E. Kohler, "Place and Practice in Field Biology," *History of Science* 40 (2002): 189–210.
2. Robert E. Kohler, "Finders, Keepers: Collecting Sciences and Collecting Practice," *History of Science* 45 (2007): 428–454.
3. William J. Turkel, *The Archive of Place: Unearthing the Pasts of the Chilcotin Plateau* (Vancouver: University of British Columbia Press, 2007).
4. Recently, some scholars have sought to redress the absence of Native American perspectives in interpretations of the mounds by integrating native traditions and oral histories. See, for example, Barbara Alice Mann, *Native Americans, Archaeologists, and the Mounds* (New York: Peter Lang, 2003). Mann has also recently helped produce an interactive virtual tour of ancient Ohio mounds in conjunction with the University of Cincinnati's Center for the Electronic Reconstruction of Historical and Archaeological Sites, available online at http://www.cerhas.uc.edu/.
5. Samuel F. Haven, *Archaeology of the United States, or, Sketches, Historical and Biographical, of the Progress of Information and Opinion Respecting the Vestiges of Antiquity in the United States*, Smithsonian Contributions to Knowledge 8 (Washington, D.C.: Smithsonian Institution, 1856), 33.
6. Some biographical information on Haven may be found online at the American Antiquarian Society website, http://www.americanantiquarian.org/Exhibitions/Portraits/samuelhaven.htm. Haven's specific interest in southern Ohio archaeology

during the 1840s is discussed in David Meltzer, "Ephraim Squier, Edwin Davis, and the Making of an Archaeological Classic," introduction to reprint edition of Ephraim G. Squier and Edwin H. Davis, *Ancient Monuments of the Mississippi Valley* (1848; Washington, D.C.: Smithsonian Institution Press 1998), 14–17.

7. Haven, *Archaeology of the United States*, 105–106.
8. George W. Knepper, *Ohio and Its People* (Kent: Kent State University Press, 1997), 111.
9. See Robert Silverberg, *Mound Builders of Ancient America: The Archaeology of a Myth* (Athens: Ohio University Press, 1968); Steven Conn, *History's Shadow: Native Americans and Historical Consciousness in the Nineteenth Century* (Chicago: University of Chicago Press, 2004), esp. chap. 4; and J. Conor Burns, "Networking Ohio Valley Archaeology in the Nineteenth Century" (Ph.D. diss., University of Toronto, 2006), esp. chap. 1.
10. Conn, *History's Shadow*.
11. R. Douglas Hurt, *The Ohio Frontier: Crucible of the Old Northwest, 1720–1830* (Bloomington: Indiana University Press, 1996), 344–347; and Robert Leslie Jones, *History of Agriculture in Ohio to 1880* (Kent, Ohio: Kent State University Press, 1983), esp. chap. 2.
12. Andrew R. L. Cayton, "The Significance of Ohio in the Early American Republic," in *The Center of a Great Empire: The Ohio Country in the Early American Republic*, ed. Andrew R. L. Cayton and Stuart D. Hobbs (Athens: Ohio University Press, 2005), 2.
13. David T. Stephens, "Population Patterns," in *A Geography of Ohio*, ed. Leonard Peacefull (Kent, Ohio: Kent State University Press, 1996), 127–145, esp. pp. 127–128.
14. Stephens, "Population Patterns," 128; and Cayton, "Significance of Ohio," 3.
15. Knepper, *Ohio and Its People*, 494; and Cayton, "Significance of Ohio," 3.
16. Cayton, "Significance of Ohio," 2–8. On the relationship between American technological progress and the creation of an earthly paradise, see also Thomas P. Hughes, *Human-Built World: How to Think About Technology and Culture* (Chicago: University of Chicago Press, 2004); and David Nye, *America as Second Creation: Technology and Narratives of New Beginnings* (Cambridge: MIT Press, 2003).
17. Thomas Tax, "The Development of American Archaeology 1800–1879" (Ph.D. diss., University of Chicago, 1973), 138; and John C. Greene, *American Science in the Age of Jefferson* (Ames: Iowa State University Press, 1984), 359–360.
18. Jones, *History of Agriculture in Ohio*, chap. 2; and Knepper, *Ohio and Its People*, 125–129.
19. Knepper, *Ohio and Its People*, 125–129.
20. Ibid., 287.
21. See esp. Ginette Aley, "Bringing About the Dawn: Agriculture, Internal Improvements, Indian Policy, and Euro-American Hegemony in the Old Northwest, 1800–1846," in *The Boundaries Between Us: Natives and Newcomers Along the Frontiers of the Old Northwest Territory, 1750–1850*, ed. Daniel P. Barr (Kent: Kent State University Press, 2006), 196–218. Recently, too, Michael Adas has explored links between technological imperatives—especially the railroad—and the "civilizing" of the West along what he describes as "not one but many frontiers." Adas, *Dominance by Design: Technological Imperatives and America's Civilizing Mission* (Cambridge: Harvard University Press, 2006), 76.

22. Aley cites a number of historians who have debunked the myth that Thomas Jefferson envisioned a more idyllic agrarian utopianism featuring the independent yeoman farmer. See Aley, "Bringing About the Dawn," 200–201.
23. Ibid., 198.
24. Caleb Atwater, "Description of the Antiquities Discovered in the State of Ohio and Other Western States," *Archaeologia Americana: Transactions and Collections of the American Antiquarian Society* 1 (1820): 105–267, on p. 121.
25. Greene, *American Science*, 358–375.
26. Atwater, "Description," 111–121 passim.
27. See especially Atwater's correlation of a three-headed vessel found in a Kentucky mound with the three chief gods of India, in Atwater, "Description," 237–241.
28. Anon. [W. W. Mather?], abstract of the "First Annual Report on the Geological Survey of the State of Ohio," *American Journal of Science and Arts* 34 (1838): 347–364.
29. Whittlesey, as quoted in abstract of the "Report on the Geological Survey," 362.
30. Ibid., 361.
31. Ibid.
32. On Hitchcock's views about the diluvial origins of mounds, see Edward Hitchcock, *Final Report on the Geology of Massachusetts* (Amherst, Mass.: Adams, 1841), 369. Also see Edwin Davis to Charles Rau, 28 May and 5 June 1884, Smithsonian Institution National Anthropological Archives, Suitland, Md. (hereafter cited as SI/NAA), MS 7065; and Marc Rothenberg, Paul Theerman, Kathleen Dorman, John Rumm, and Deborah Jeffries, eds., *The Papers of Joseph Henry*, vol. 7, *January 1847–December 1849: The Smithsonian Years* (Washington, D.C., Smithsonian Institution Press: 1996), 59n7, 510n1.
33. Terry Barnhart, "Of Mounds and Men: The Early Anthropological Career of Ephraim George Squier" (Ph.D. diss., Miami University of Ohio, 1989), 33–36.
34. Davis to Rau, 28 May and 5 June 1884, SI/NAA, MS 7065.
35. Burns, "Networking Ohio Valley Archaeology," chaps. 1–2.
36. Meltzer, "Making of an Archaeological Classic"; and J. Conor Burns, "Ethnological Diversity in a New Light: Daniel Wilson's 1858 Critique of North American Ethnology," paper delivered at the History of Science Society meeting, November 2006, Vancouver, British Columbia.
37. Meltzer, "Making of an Archaeological Classic," 21–22.
38. On the collaboration between Squier and Davis and the Smithsonian, see Meltzer, "Making of an Archaeological Classic"; Barnhart, "Of Mounds and Men"; Terry Barnhart, *Ephraim George Squier and the Development of American Anthropology* (Lincoln: University of Nebraska Press, 2005); and Tax, "Development of American Archaeology." Correspondence pertaining to the collaboration is included in Rothenberg and others, *Papers of Joseph Henry*, vol. 7.
39. Barnhart, "Of Mounds and Men," 160–165; and Meltzer, "Making of an Archaeological Classic," 51–54.
40. Squier and Davis, *Ancient Monuments of the Mississippi Valley*, xxxix.
41. Henry Howe, *Historical Collections of Ohio* (Cincinnati: C. J. Krehbiel and Co. for the State of Ohio, 1904), 2:491, 495–498. The complete subtitle of this useful source deserves to be cited in full: "An Encyclopedia of the State: History both general and local, geography with descriptions of its counties, cities, and villages, its agricultural,

manufacturing, mining, and business development, sketches of eminent and interesting characters, etc., with notes of a tour over it in 1886. Illustrated by about 700 engravings. Contrasting the Ohio of 1846 with 1886–90. From drawings by the author in 1846 and photographs taken solely for it in 1886, 1887, 1888, 1889, and 1890, of cities and chief towns, public buildings, historic localities, monuments, curiosities, antiquities, portraits, maps, etc."

42. Squier and Davis, *Ancient Monuments*, xxxix.
43. Charles Whittlesey, *Ancient Works in Ohio*, Smithsonian Contributions to Knowledge 3 (Washington, D.C.: Smithsonian Institution, 1851); and Whittlesey, *Ancient Mining on Lake Superior*, Smithsonian Contributions to Knowledge 13 (Washington, D.C.: Smithsonian Institution, 1863).
44. See, for example, Joseph Henry, "Report of the Secretary," *Smithsonian Annual Report for 1853*, 170–171. This entry is characteristic of those to appear throughout the 1850s to 1870s. On the Smithsonian's correspondent community in general, see E. F. Rivinius and E. M. Youssef, *Spencer Baird of the Smithsonian* (Washington, D.C.: Smithsonian Institution Press, 1992); and Daniel Goldstein, "'Yours for Science': The Smithsonian Institution's Correspondents and the Shape of Scientific Community in Nineteenth-Century America," *Isis* 85 (1994): 573–599. Certainly, Spencer Baird deserves at least equal credit for cultivating the Smithsonian's body of correspondent field-workers.
45. Joseph Henry, "Report of the Secretary," *Smithsonian Annual Report for 1865* (1872), 14–15.
46. The body of secondary literature addressing the rise of evolutionist anthropology and the unity and antiquity questions is too large to cite in full here, but see, for example: Donald Grayson, *The Establishment of Human Antiquity* (New York: Academic Press, 1983); Alice Beck Kehoe, *The Land of Prehistory: A Critical History of American Archaeology* (New York: Routledge, 1998); Adam Kuper, "Anthropology," in *The Cambridge History of Science*, vol. 7, *The Modern Social Sciences*, ed. Theodore M. Porter and Dorothy Ross (Cambridge: Cambridge University Press, 2003), 354–378; David J. Meltzer, "The Antiquity of Man and the Development of American Archaeology," in *Advances in Archaeological Method and Theory*, vol. 6, ed. Michael B. Schiffer (New York: Academic Press, 1983), 1–51; G. Blair Nelson, "'Men Before Adam!': American Debates over the Unity and Antiquity of Humanity," in *When Science and Christianity Meet*, ed. David C. Lindberg and Ronald L. Numbers (Chicago: University of Chicago Press, 2003), 161–181; George W. Stocking Jr., *Victorian Anthropology* (New York: Free Press, 1987); Stocking, "Paradigmatic Traditions in the History of Anthropology," in *Companion to the History of Modern Science*, ed. R. C. Olby, G. N. Cantor, J. R. R. Christie, and M. J. S. Hodge (London: Routledge, 1990), 712–727; and A. Bowdoin Van Riper, *Men Among the Mammoths: Victorian Science and the Discovery of Prehistory* (Chicago: University of Chicago Press, 1993).
47. On the formation of the Peabody Museum, see Curtis Hinsley, "The Museum Origins of Harvard Anthropology, 1866–1915," in *Science at Harvard University: Historical Perspectives*, ed. Clark Elliott and Margaret Rossiter (Bethlehem, Pa.: Lehigh University Press, 1992), 121–145; Hinsley, "Frederic Ward Putnam," in *Encyclopedia of Archaeology: The Great Archaeologists*, vol. 1, ed. Tim Murray (Santa Barbara, Calif.: ABC-CLIO, 1999), 141–153; Tax, "Development of American

Anthropology," 307–316; Frederic W. Putnam, "The Peabody Museum of American Archaeology and Ethnology," *Proceedings of the American Antiquarian Society*, n.s., 6 (1890): 180–190; "Mr. George Peabody's Recent Gifts to Science," *American Journal of Science and Arts*, 2nd ser., 43 (1867): 131–133; "Death of George Peabody," *American Journal of Science and Arts*, 2nd ser., 48 (1869): 442–445; Joseph Henry, "Report of the Secretary," *Smithsonian Annual Report for 1868*, 26–27.

48. The fullest available account of the early history of the Bureau is found in Curtis Hinsley, *The Smithsonian and the American Indian: Making a Moral Anthropology in Victorian America* (Washington, D.C.: Smithsonian Institution Press, 1994).
49. John Wesley Powell, "Introduction," *Twelfth Annual Report of the Bureau of Ethnology for the Years 1890–1* (1894), xl–xli, esp. p. xli; and Stephen Williams, *Fantastic Archaeology: The Wild Side of North American Prehistory* (Philadelphia: University of Pennsylvania Press, 1991), 64–65.
50. Burns, "Networking Ohio Valley Archaeology," chaps. 5–6.
51. On the development of correspondence networks at both Smithsonian and the Peabody, see Rivinius and Youssef, *Spencer Baird of the Smithsonian*; Goldstein, "Yours for Science"; Hinsley, "Museum Origins of Harvard Anthropology"; and Hinsley, "From Shell-Heaps to Stelae: Early Anthropology at the Peabody Museum," in *Objects and Others: Essays on Museums and Material Culture*, ed. George W. Stocking Jr. (Madison: University of Wisconsin Press, 1985), 49–74.
52. When the Peabody Museum began operations in the 1860s, Joseph Henry optimistically called for a "union of effort" between it and the Smithsonian "to extensively examine the [ancient] monuments and collect all the relics, to illustrate as fully as possible the archaeology and ethnology of the American continent" (*Smithsonian Annual Report for 1868*, 26). In a broad sense, the two institutions did subsequently function in a cooperative spirit, jointly participating in organized government expeditions and exchanging specimens. When it came to these mound initiatives, however, competition became a driving force.
53. I develop this particular argument and its implications for understanding processes of institutional centralization in much greater detail elsewhere. See J. Conor Burns, "Networking Ohio Valley Archaeology in the 1880s: The Social Dynamics of Peabody and Smithsonian Centralization," *Histories of Anthropology Annual* 4 (2008): 1–33.
54. See, for instance, J. P. MacLean, "Ancient Remains in Ohio," *Smithsonian Annual Report for 1885*, 893–900. In each of his descriptions of a number of ancient works in southern Ohio, MacLean routinely includes information about the amount of plowing done over each site. Only once could he report that a mound had not been disturbed by the plow.
55. Howe, *Historical Collections*, 2:491. According to Howe's census-based statistics for 1887, Ross Country comprised 650 total square miles, of which 119,709 acres were under cultivation and 107,699 in use for pasture.
56. On the background and early history of the OSAHS, see Terry Barnhart, "In Search of the Mound Builders: The State Archaeological Association of Ohio, 1875–1885," *Ohio History* 107 (1998): 125–170.
57. See George F. Wright, "Importance of the Study of Archaeology in Ohio," *Ohio Archaeological and Historical Publications* 1 (1887–88): 54–59. In this piece, Wright

provided a brief introduction to a collection of excerpts of Putnam's writings on the subject, from personal correspondence as well as published Peabody Museum reports covering the previous few years. Wright was a prominent Oberlin College geologist and mainstay of the OSAHS, and one of the many figures with whom Putnam corresponded.

58. Putnam, as quoted in Wright, "Importance," 58. See also Putnam's letter to the *Cincinnati Post* dated June 4, 1887, reprinted as "The Serpent Mound Saved," *Ohio Archaeological and Historical Publications* 1 (1887–88): 184–187. Putnam's concerns for the agricultural destruction of mounds began earlier in the decade.
59. Frederic Ward Putnam, "The Serpent Mound of Ohio," *Century Magazine* 39 (1889–90): 871–888.
60. Frederic Ward Putnam, "Explorations of the Harness Mounds in the Scioto Valley, Ohio," *Eighteenth Annual Report of the Peabody Museum* (1885), reprinted in *The Archaeological Reports of Frederic Ward Putnam* (New York: AMS Press, 1973), 224–227; and Frederic Ward Putnam to Esther Orne Clarke Putnam, 30 May 1887, Harvard University Archives, Unit HUG1717.2.1 Putnam Papers, General Correspondence, 1851–1947 (1881–1890), Box 7 ("M–Z"), Folder "P." See also Frederic Ward Putnam, "Archaeological Explorations in Ohio and Wisconsin," *Seventeenth Annual Report of the Peabody Museum* (1884), reprinted in *The Archaeological Reports of Frederic Ward Putnam*, 209–221.
61. James D. Middleton to Cyrus Thomas, 1 July 1888, SI/NAA, Division of Mounds Exploration Records, 1881–1889, MS 2400, Folder "J. D. Middleton, Ross and Licking Counties, 1888–9." See also any other of Middleton's letters in this folder.
62. This became a source of headache for Thomas, who now had to include not just re-surveying but also explicitly correcting Squier and Davis's errors as part of the Mound Exploring Division's duties. Thomas addressed the issue in a special Bureau publication. See Cyrus Thomas, *The Circular, Square, and Octagonal Earthworks of Ohio* (Washington, D.C.: Government Printing Office, 1889).
63. Curtis Hinsley discusses Metz as part of Putnam's "correspondence school." See Hinsley, "Museum Origins of Harvard Anthropology."
64. Burns, "Networking Ohio Valley Archaeology in the 1880s," 8–9.
65. Putnam, "Archaeological Explorations in Ohio and Wisconsin," 212. Many of Putnam's reports included references to having gained right of exploration from farmers. See, for instance, Putnam, "Explorations of the Harness Mounds"; and Putnam, "Explorations in the Little Miami River Valley," *Twentieth Annual Report of the Peabody Museum* (1887), reprinted in *The Archaeological Reports of Frederic Ward Putnam*, 235–250.
66. Metz to Putnam, 25 November 1888, Peabody Museum Archives, Accession File (PM/AF) 88-55-1, "Expedition to Ohio-Serpent Mound."
67. Burns, "Networking Ohio Valley Archaeology," chap. 6.
68. Putnam, "Explorations of the Harness Mounds," 223–224.
69. See, for instance, Spencer Baird to Cyrus Thomas, 29 July and 4 August 1884, SI/NAA, Division of Mounds Exploration Records, 1881–1889, MS 2400, Box 8.
70. Smith to Baird, 22 August 1884; Smith to Powell, 15 September 1884; and Smith to Thomas, 29 September 1884, all from SI/NAA, Division of Mounds Exploration Records, 1881–1889, MS 2400, Box 5, Charles M. Smith Correspondence, 1884–1887.

71. Charles Smith, "A Sketch of Flint Ridge, Licking County, Ohio," *Smithsonian Annual Report for 1884*, 851–873.
72. Mark Barrow, "The Specimen Dealer: Entrepreneurial Natural History in America's Gilded Age," *Journal of the History of Biology* 33 (2000): 493–534.
73. One of Putnam's field correspondents commented at the end of the decade on the "unfortunate price" paid by the Smithsonian for one collection in particular, and despaired of the trouble this would cause as collectors would quickly realize how much money they could get. Cresson also worried about the impact on archaeological collecting as a result of Putnam's preparations for the Anthropology Department of the 1893 Chicago World's Columbian Exposition. As head of this department, Putnam's work here included organizing a massive archaeological exhibit, something Cresson believed had opened up a "relic boom" among entrepreneurial collectors. Some of these collectors, Cresson believed, employed unscrupulous methods in the name of making a profit. See Harlan Cresson to Putnam, 29 December 1890, Peabody Museum Archives, Director Records of Frederic Ward Putnam, 89-15, Box 11 "1891 A–F," Folder "UAV 677.38." Another related concern was a rising number of fake or forged specimens being passed off as authentic. See Putnam, "Archaeological Frauds," *Science* 1 (1883): 99.
74. See Terry Barnhart, "An American Menagerie: The Cabinet of Squier and Davis," *Timeline* 2 (December 1985–January 1986): 2–17.
75. Smith to Thomas, 28 January 1885, SI/NAA, Division of Mounds Exploration Records, 1881–1889, MS 2400, Box 5, Charles M. Smith Correspondence, 1884–1887.
76. Moorehead to Baird, 3 March 1885, SI/NAA, Division of Mounds Exploration Records, 1881–1889, MS 2400, Box 6, Folder "Warren K. Moorehead, Licking County, 1885–6."
77. Hampton to Powell, 6 May 1885, SI/NAA, Division of Mounds Exploration Records, 1881–1889, MS 2400, Box 6, Folder "W.C. Hampton, Hardin County, 1885."
78. Sarah DeWitt Dunlap to Frederic Ward Putnam, 23 December 1889, Peabody Museum Accession Files, Unit 88-55, Folder 1.
79. Dunlap to Putnam, 23 December 1889, Peabody Museum Accession Files, Unit 88-55, Folder 1. William Hamilton was Dunlap's maternal grandfather, and she knew that he owned the farm at the time of her parents' marriage in 1820.
80. See, for example, Ephraim Squier, *The Serpent Symbol, and the Reciprocal Principles of Nature in America* (New York: George P. Putnam, 1851); Stephen Peet, "The Religious Character of the Emblematic Mounds," *American Antiquarian and Oriental Journal* 6 (1883–84): 393–411; Peet, "The Serpent Symbol in America," *American Antiquarian and Oriental Journal* 8 (1885–86): 197–221; Peet, "The Great Serpent and Other Effigies," *American Antiquarian* 12 (1890): 211–28; and Putnam, "Serpent Mound of Ohio."
81. A transcript of Putnam's address detailing the visit may be found in *Proceedings of the American Antiquarian Society*, n.s., 3 (October 1883–April 1885): 4–20.
82. John P. MacLean, "The Great Serpent Mound," *American Antiquarian and Oriental Journal* 7 (1885): 44–47; Anon., "The Serpent Mound Saved," *Ohio Archaeological and Historical Quarterly* 1 (1887–88): 187–190; and Otis Mason, "The Great Serpent Mound," *American Naturalist* 21 (1887): 786–787.

83. Putnam, "Serpent Mound of Ohio," 871–873.
84. Ibid., 878–888.
85. Thomas built a case for a connection between the Ohio moundbuilders and the Cherokees. See Cyrus Thomas, *The Problem of the Ohio Mounds* (Washington, D.C.: Government Printing Office, 1889). See also Thomas, "Who Were the Mound-Builders," *American Antiquarian* 6 (1884): 90–99; and Powell, "Introduction," xl–xli.
86. Other forms of development and construction certainly still presented problems, but a further serious concern was an increasing amount of site looting, fuelled largely by the growing market for antiquities. See Putnam, "Archaeological Frauds," 99; and Stephen Peet, "The Destruction of Mounds," *American Antiquarian* 6 (1883–84): 41.
87. Curtis Hinsley has used the term "salvage ethnology" to characterize the general ethos of nineteenth-century U.S. ethnographic and linguistic anthropology, in *Smithsonian and the American Indian*, 22–23.

Four

Stuart McCook

Managing Monocultures

Coffee, the Coffee Rust, and the Science of Working Landscapes

Many of the field sciences aim to study a nature that is pristine—at least rhetorically pristine. Field scientists would—quite literally—climb the highest mountains and plumb the deepest oceans in the quest of pure, pristine nature. They would travel great distances from cities and settled areas, deep into the forests in pursuit of an undisturbed field to study, to examine plants and animals in their "wild" habitat. If some traces of human activity were present, field scientists developed conceptual tools that allowed them to treat landscapes *as if* they were wild. More recently, field scientists interested in conservation have begun to explicitly include anthropogenic factors in their study of topics such as wildlife habitat. But even here, these anthropogenic forces are understood as external forces impinging on a wild or pristine nature. Few of the field sciences take anthropogenic landscapes or environments as the explicit *object* of their study—as something to be explained, rather than explained away.

As far as field scientists were concerned, anthropogenic landscapes—farms, managed forests, and so forth—seemed to belong to a lesser nature, one that was less scientifically interesting than wild nature.[1] Working landscapes—domesticated forests, cultivated fields, working maritime environments—did not receive the same scientific attention as did purportedly wild or natural landscapes. This attitude began to change somewhat during the late eighteenth and early nineteenth centuries. Naturalists working in the tropics, for example, explored the connections between deforestation and local climate change. In Europe, the pioneering German chemist Justus von Liebig (1803–1873) developed techniques for conducting systematic chemical analyses of agricultural soils. He formulated a scientific approach to studying agricultural landscapes that was widely emulated throughout Europe and in North America.[2] This new scientific interest in working landscapes—and the emergence of new applied disciplines such as agricultural chemistry and plant pathology—emerged because of unprecedented and historically specific environmental changes in working landscapes. The science of agricultural chemistry, for example, emerged at least in part because of declining soil fertility in central Europe.

The scientific study of working landscapes differed from that of wild landscapes in three important ways. First, the *structure* of working landscapes was often simpler—or at least more specialized—than that of wild landscapes. Working landscapes can be defined as landscapes that have been transformed and managed to meet the needs of a society. The nature of these transformations varies considerably from one society to the next, and from one period to the next. In the nineteenth and twentieth centuries, for example, Europeans (and their diasporas) greatly expanded the practice of monoculture at home and in their colonies. The ecology of working landscapes often functioned differently than that of wild landscapes—both by design and by accident. Phenomena that were rare in wild landscapes, such as diseases and pests, were much more common in working landscapes. Second, the *object of study* in working landscapes included *people*, as well as the usual plants and animals. When studying working landscapes, field scientists had to account for human intentions and actions—something that they rarely had to do for wild landscapes. Third, the *production of knowledge* about working landscapes often involved these local actors. Scholars often distinguish between "scientific" knowledge produced by scientists and "local" knowledge produced by people who live and work in the field. The two are often contrasted—the high modernist, European science standing in sharp distinction to the local, traditional techne practiced by peasants or indigenous peoples.[3] In practice, the distinction between these two kinds of knowledge is not always clear. European scientists often appropriated and incorporated local knowledge into their theories; the locals just as often selectively appropriated and incorporated scientific ideas and practices into their own knowledge systems.

In working landscapes, therefore, the interplay between nature(s) and knowledge(s) was much more dynamic and volatile than in wild landscapes. One important instance of this volatility is the wave of acute crop pandemics and pest infestations that spread across the globe during the nineteenth and twentieth centuries. The potato blight fungus (*Phythopthora infestans*) traveled from South America through North America to Europe, contributing to the Irish Potato Famine of the 1840s. French vineyards were stricken with an epidemic of powdery mildew (*Oidium tuckeri*) and with an infestation of the phylloxera insect, which together threatened the entire French wine industry with collapse.[4] These pandemics reflected historically specific changes in global trade and agriculture. They also highlighted just how poorly scientists—or anyone else—understood the workings of agricultural landscapes. Farmers and scientists alike struggled to understand and control these epidemics; both groups shaped the changing ideas of plant disease during this period.

In the nineteenth century, explanations for crop diseases went through broadly similar transformations as did explanations for human diseases. In the earlier part of the century, ecological explanations of disease predominated. These focused on the role of soil, climate, and other predisposing factors in shaping the crop epidemic. In this model, fungi and other such signs of "decay" were interpreted as *consequences* of the epidemic, rather than their cause. Later in the century, pathogenic models of disease began to prevail in scientific and popular accounts of crop disease. Pathogenic models stressed the role of disease-causing organisms, especially fungi, in causing plant epidemics. The ecological and pathogenic models of disease were not polar opposites, but rather points on a spectrum. Not only were the *theories* of crop disease changing; the biological nature of the epidemics themselves was also changing. In short, in the nineteenth century, the *ontology* of these crop epidemics was changing almost as quickly as the *ideas* about them. There was a constant interplay between the (dynamic) landscapes and the (dynamic) ideas of disease.

One of the most important of these epidemics was the global epidemic of coffee rust. The epidemic was first detected in Ceylon in 1869; over the next century it spread to virtually every coffee-growing region in the world. Most plant pathology textbooks focus on one brief, two-year interval in the epidemic, from 1880 to 1882. During these two years the British botanist Harry Marshall Ward (1854–1906) conducted pioneering research on the disease. He was among the first scientists to demonstrate a connection between crop epidemics and particular agricultural practices. For this insight, Ward's work in Ceylon has a rightful place in the canon of plant pathology. Ward's researches in Ceylon have become canonical stories recounted in plant pathology and mycology textbooks—even though Ward offered no practical means to control the coffee rust epidemic, and even though he never again studied coffee rust during his long career. Like many canonical stories, however, the case of Harry Marshall Ward and the coffee rust masks more complex—and often more interesting—stories. By shifting focus from Ward to the coffee rust epidemic itself, we can learn a lot more about how agricultural knowledge was produced and circulated in the late nineteenth and early twentieth centuries, and about how this knowledge was shaped by changes in the working landscapes.[5]

Agricultural Knowledge and Scientific Knowledge Until 1850

In the mid-nineteenth century, Ceylon was one of Great Britain's most dynamic tropical colonies. The British began conquering and colonizing the island in earnest during the 1820s. In the following decades, they conquered the interior highlands in the Kingdom of Kandy. These landscapes proved to be ideal for growing coffee, and British planters began cultivating the crop on a large scale. The island was also home to what was arguably the British Empire's most important tropical botanical garden: the Royal Botanic Garden at Peradeniya, in the center of the island, surrounded by coffee groves. Nonetheless, the garden's scientific work and the planters' agricultural work remained almost entirely separate enterprises. The garden's naturalists devoted most of their time to making inventories of the island's flora and fauna, and to acclimatizing exotic crops. They had little to offer the coffee farmers; coffee was an exotic crop, but it had been introduced to Ceylon and domesticated there several centuries before the British arrived. The island's British coffee farmers looked to one another, rather than to the garden's scientists, for horticultural and agricultural advice. New planters learned through a system of apprenticeship known as "creeping."

Other coffee planters migrated from elsewhere in the empire—particularly from the Caribbean, where slavery had recently been abolished. Experienced and new planters alike drew on a small but growing body of agricultural literature on coffee cultivation, especially P. J. Laborie's *The Coffee Planter of Saint Domingo*. This treatise on coffee growing, based on the author's experience operating a plantation in the French Caribbean colony of Saint-Domingue, was initially published in 1797. For the next half century, it was the bible of European coffee growers around the world. On small farms, coffee was commonly grown as a garden crop. On large estates, planters such as Laborie promoted intensive agriculture—with landscapes devoted almost exclusively to coffee, which was to be planted in carefully calculated, orderly rows. This form of cultivation would maximize production and also allow overseers to keep an eye on their labor. It also satisfied an aesthetic ideal of orderly, rational agriculture.[6]

The coffee rust epidemic broke out at a moment when both scientific and popular understandings of health and disease were changing from a holistic, environmental model to a pathogenic model. Early naturalists who studied crop diseases concluded that disease was the result of a "disturbed condition" of the plant, possibly brought about by a combination of unfavorable conditions in the soil, the climate, and the host plant. The holistic model of crop disease made observational sense; for example, epidemics such as the potato blight were clearly associated with unusually wet seasons. Naturalists often observed fungi, mildews, and other organisms on diseased plants. But these organisms were understood to be a *consequence* of the disease, rather than its cause. In this view, such organisms were produced by spontaneous generation (or heterogenesis), from "abnormal sap or tissue."

This view was not, however, necessarily universal. As early as the mid-eighteenth century, French botanists such as Mathieu Tillet and Isaac-Bénedict Prévost had begun to argue that rusts and smuts were, in fact, parasitic plants that *caused* disease. The debate over fungal pathogenicity became acute during the potato blight of the 1840s. The connections between the fungus and the crop failures were far from clear, and naturalists debated the nature of the blight quite vigorously into the 1850s. The German naturalist Anton de Bary (1831–1888) developed a new experimental method for studying crop diseases. His research, conducted at the Universities of Freiburg, Halle, and Strasbourg, lent support to the pathogenic model of plant disease. De Bary took an innovative approach to studying fungi. Through meticulous research in the laboratory and the field, he tried to reconstruct the entire life cycle of the fungus from spore to mature organism. He cultivated spores both in the laboratory and on plants, and tried to reproduce disease by systematically inoculating healthy plants with spores of the fungus. Between 1853 and 1863, he conducted extensive research on the rusts and smuts of wheat, and on the potato blight. He produced convincing evidence that the fungi were independent organisms, that they had a life cycle, and that they were the cause of plant diseases, rather than their consequence. Other scientists in Europe and elsewhere quickly extended de Bary's researches and insights to other crops. De Bary himself was primarily interested in understanding the nature of fungi; he had little interest in studying methods of disease control. These tasks were left to other researchers.[7] Even so, de Bary's model was not universally accepted, either by other scientists or by farmers.

This was the state of plant pathology in 1869, the year a farmer in central Ceylon first noticed rust-colored spots growing on the leaves of his coffee trees, followed by a massive defoliation. The farmer took some of the leaves to G. H. K. Thwaites, the director of the Peradeniya botanical garden. Through the 1850s and 1860s, Thwaites had been studying the fungi of Ceylon, and he had never encountered anything like what he found on the infected coffee leaves. He forwarded some infected leaves to his correspondent in Britain, Miles Joseph Berkeley (1803–1889). At the time, Berkeley was the country's leading mycologist. In many respects, Berkeley's career was more typical of an older generation of naturalists. He was a university graduate but did not work in a university. Rather, he earned his living as a country vicar; he rose to scientific eminence in his spare time. His research focused on the morphology of the fungi; he did not conduct experimental work the way that de Bary and the new generation of botanists did. Nonetheless, Berkeley was also one of the earliest—and most vocal—adherents to the theory of fungal pathogenicity, based on his field research on the potato blight in the 1840s. In a widely read paper published in 1846, he concluded that "the decay [in the potato plant] is a consequence of the presence of the mould, and not the mould of the decay."[8] Unlike many naturalists at the time, Berkeley saw the fungus as a pathogen—as the cause of the disease rather than as a symptom of it. The fungus had first been collected in the field some time early in 1869; by July or August of that year, it had reached Berkeley, who forwarded it on to the microscopist C. E. Broome. On the first of September Broome sent a series of preliminary drawings of the fungus, based on his microscopical observations, back to Berkeley.

Broome seemed puzzled by what he had found. "I have so little to say about your fungus of coffee," he wrote, "that I had better leave it to you to complete your reply to Thwaites."[9] As it turns out, Broome was right to be confused about the fungus: it appeared to be a completely new species. "The most curious circumstance," observed Berkeley, "is that amongst more than a thousand species of Fungi received from Ceylon, this does not occur." The coffee leaf fungus looked like no other, and Berkeley concluded that "it is not only quite new, but with difficulty referable to any recognized section of the fungi." The fungus was so different from known fungi that Berkeley concluded that it represented not only a new species, but also a new genus. Berkeley named the genus "Hemileia" (for half smooth, reflecting the shape of the spores). He named the species, aptly as it turned out, "vastatrix"—the Latin word for "devastator" or "destroyer." He published the first drawings and descriptions of the new fungus in the *Gardeners' Chronicle* of November 6, 1869. Only six months had passed between the discovery of a new fungus in a remote coffee farm in the interior of Ceylon, its description in an equally remote vicarage in Britain, and its publication in one of Europe's leading botanical journals.[10]

Berkeley's identification of the fungus marked the beginning of scientific research into the coffee rust epidemic, but many scientific and agricultural problems remained. The core scientific problem was to fully describe the fungus's life cycle, and to account for how it behaved in the field. This problem was particularly acute in Ceylon. The island's humid climate, fueled by two annual monsoon seasons, was ideal for the growth and propagation of fungi of all kinds. Hundreds, if not thousands, of species of fungi could be found in the island's coffee farms. During the 1870s, this problem seemed almost insurmountable—largely because of the great distances between the experts and the field. The handful of scientists with the requisite expertise in fungi lived in Europe. They could not observe the behavior of the live fungus in the field, and so had little hope of understanding the fungus's life cycle. Without an understanding of this cycle, as far as scientists were concerned, it would be impossible to develop effective measures to control the disease. Berkeley argued that the disease would be difficult to control since "the fungus is confined to the underside of the leaves, and the mycelium is not superficial." Still, he recommended that planters try to use sulfur to control the outbreak, either by spraying or syringing it onto the infected areas of the leaves.[11] Few coffee planters experimented with sulfur as a means of controlling the disease; chemical controls meant heavy investments in chemicals, technology, and labor. In the early 1870s, however, coffee prices had entered a slump as Brazilian coffee flooded the global markets.

More important, during the 1870s many coffee planters in Ceylon did not accept the pathogenic model of disease. They continued to invoke physiological and environmental explanations for the epidemic. Some argued that the disease was caused by a "poisoning of the juices" of the coffee tree. They suggested that the disease was the result of poor cultivation, of inadequate manuring, or of climatic disturbances. Some planters argued that the disease was just temporary, and that sooner or later it would "wear itself out"—as earlier outbreaks of diseases and pests appear to have done.[12] In fact, some of these criticisms seemed to be borne out in the field. In the early years of the outbreak, coffee growers who applied manure to their farms found that coffee yields recovered. They concluded, therefore, that the disease was essentially caused by soil exhaustion, and that manuring could "cure" it. In this period, as later, there was a strong moral undertone to the planters' response to the disease; a plantation afflicted with the epidemic must be a plantation that was somehow poorly cultivated.

These observations reflect the epidemic's seemingly erratic behavior, which confounded many observers, including Thwaites. Coffee trees that had been completely defoliated one year—left dry and brittle—seemed to recover almost completely the next year, producing what appeared to be a full crop of berries. Still, even Thwaites became gradually more pessimistic through the early 1870s. He vigorously argued that the fungus was the cause of the disease. "There can be no question," he claimed, "that the fungus is communicated from coffee plant to coffee plant through dissemination of the spores, and that this may be conveyed by the wind, or by streams of water, or by animals of many kinds moving from place to place." Like many coffee planters, Thwaites at first hoped that the epidemic might just run its course and disappear.[13] But this hope gradually faded, and by 1874 he explicitly contradicted this view: "Some persons would seem to be under the impression that in the course of time the leaf disease will wear itself out and entirely disappear, but it is difficult to see how this will happen whilst Coffee trees remain for it to subsist upon."[14]

As each season passed, it gradually became apparent that the epidemic followed a cyclic pattern, which echoed the biennial cycle of cultivated coffee. When coffee is cultivated intensively, it tends to bear heavily one year and lightly the following year. The coffee trees spend so much of their energy producing coffee berries one year that they are exhausted and need a year to recover. Thwaites and the coffee planters observed that the rust tended to exacerbate this biennial cycle. In good years, coffee production remained high, although perhaps not as high as it would have without the disease. In the off years, the defoliated trees produced even fewer berries than they would have in "normal" off years. The leaves provide vital nutrition both to the plant and to the berry. This helped explain how manure apparently cured the disease: it provided coffee trees with some additional nutrients that helped offset the nutrient losses caused by defoliation. Nonetheless, Thwaites and others soon dismissed the notion that manure was a cure. They found that heavy manuring could temporarily offset some of the losses caused by the epidemic, but its benefits disappeared after a season or two.[15] Thwaites concluded that the rust attacks, "occurring periodically, at length seriously affect the health of the tree which, if old and ill-cultivated, becomes of little or no value as a crop producer."[16]

The New Botany and the Rust Epidemic

Fearing the growing losses from the epidemic, and frustrated by the slow pace of scientific research, in 1879 coffee planters in Ceylon lobbied the colonial government to hire a scientist who would devote himself exclusively to studying the epidemic. The planter G. A. Talbot wrote that the planters needed "a scientific man to make what researches he can and to give us information from a scientific point of view, so as to help us carry on the experiments. From a practical point of view we know our business, but from a scientific point of view we can get valuable assistance, by investigations with the microscope for instance."[17] The government duly appointed a young naturalist named Daniel Morris to the post. Morris had been trained in the New Botany of the mid-nineteenth century, studying at the Normal School of Science in South Kensington under William Thiselton-Dyer. Thiselton-Dyer (1843–1928) was the assistant director of the Royal Botanic Gardens at Kew, had a strong interest in agriculture and applied botany, and enthusiastically promoted the New Botany in Britain. The New Botany was heavily influenced by the experimental botany being developed by Julius von Sachs (1832–1897) and Anton de Bary in Germany. It focused on the study of plant physiology and plant health, and on the behavior of living plants in the field. This stood in contrast to the older, taxonomic botany, which used the descriptive and taxonomic methods of traditional botany. By focusing on plant physiology, the New Botany also addressed issues that were more directly relevant to agricultural questions. As assistant director at Kew, Thiselton-Dyer promoted a new role for the institution as an imperial center for research into economic botany and agriculture. Through Thiselton-Dyer's influence, Morris had been posted to Peradeniya in 1877. In 1879, Morris was seconded from the garden to work exclusively on the rust.[18]

Through 1879 and 1880, Morris conducted detailed researches on the life cycle of the rust. He also experimented with different forms of chemical control. Reflecting his more holistic approach to the coffee ecosystem, however, he argued that efforts at chemical control would be futile without accompanying cultural control. In particular, he insisted that old and "ill-cultivated" infected coffee farms should be uprooted. "But, if in addition to what has already been done for them," Morris cautioned, "[coffee planters] ask for further aid from Government it is only reasonable to ask that in return *they give effect to the repeated demands made for the removal of all diseased and worthless trees*, and when this first and most necessary step it taken that they *unite in the general application of suitable remedies*."[19] In fact, Morris recommended that the total area of coffee under cultivation be *reduced* and cultivated more carefully.[20] Neither of these options was particularly appealing to the island's coffee planters.[21]

Still, coffee growers in Ceylon responded warmly to the Morris's studies of using a lime-sulfur mixture to control the disease. Morris's combination of Grigson's mixture and cultural control seemed to offer planters a solution that was both economically and agronomically viable. "The death knell of the coffee leaf-fungus," said an article in the *Indian Agriculturist* in May 1879, "is, we think, sounding, and this great scourge of the coffee planter has, after so many years of unchecked reign, at last been firmly grappled with."[22] Scientists also celebrated Morris's work as a vindication for applied scientific research in agriculture—which had been somewhat suspect on the island up until that point. "Scientific men, teachers, Coffee-planters, all alike," wrote one of the editors of the *Gardeners' Chronicle*, "may well feel some complacency when they look back at the progress made in ten—we may say, for all practical purposes, in three years. It is a justification for the existence of scientific committees, scientific lectures, and practical experiments, and we are heartily pleased to see that Ceylon planters fully appreciate the import of what has been done."[23] Planters and scientists alike celebrated the promised vindication of Ceylon's coffee offered by Morris's research—even though the rust continued to spread through the island's coffee plantations. Optimism about Morris's treatments ran high, and as one article put it, "We can pardon a tendency in interested parties to halloo before they are quite safe out of the wood."[24] Ceylon's coffee planters appeared to embrace the New Botany as epitomized by Morris. "Mr. Morris has been in this country over a year, Dr. Thwaites more than thirty," observed one coffee planter. "Who has told us the most about leaf disease?"[25] The coffee planters reacted in dismay when they found out that Morris was being transferred to the Jamaica botanical garden. In contrast, the planters welcomed Thwaites's retirement from the Peradeniya, since they felt that his pessimistic predictions of the rust's severity "were harmful to the colony's financial reputation."[26] They criticized Kew's choice for a replacement, Henry Trimen, and requested that the government replace Morris with another cryptogamist.

In response to demands from the planters and from Ceylon's governor, the British government sent another acolyte of Thiselton-Dyer's to replace Morris. Unlike Morris, however, Harry Marshall Ward had been explicitly trained in the biology of fungi. He had studied botany under Thiselton-Dyer at the science school in South Kensington, after which he won a scholarship to Cambridge, where he studied for the Natural History Tripos. He graduated with a first-class degree in 1879, and then went to Würzburg, Germany, to work with the botanist Julius von Sachs. Both at Cambridge and Würzburg, Ward's researched focused on the physiology of plants. When he returned from Germany, he worked at the Jodrell Laboratory at Kew, the newly founded center for experimental plant biology. Ward was the obvious choice for the job when the Colonial Office asked Kew to recommend a suitable candidate to replace Morris. Thiselton-Dyer and Hooker agreed that it was important to send out a researcher "versed in modern methods of microscopical investigation." They recommended that Ward be sent out on a two-year contract, renewable for a third. This, they felt, was enough time for him to study the disease and attempt to find a control for it.[27]

Even more than Morris, Ward epitomized the ideas of the New Botany. His research in Ceylon moved along two main lines. The first line was experimental. He conducted carefully controlled experiments on coffee plants in the botanical garden, and also on plants cultivated in small, portable greenhouses, known as Wardian cases. In February 1880, shortly after he arrived on the island, he planted healthy coffee seedlings from Jamaica at the botanical garden, in preparation for experimental work. He allowed the leaves to become infected, and then studied the rust "grains." He concluded that these grains were in fact spores, which could germinate. The second line of research involved the systematic, comparative study of working coffee plantations across the island. On March 16, 1880, he began a tour through the coffee districts, to assess the disease's impact and to collect specimens for analysis. He systematically collected samples from a wide range of areas, elevations, and soil conditions. In addition, he also began systematically observing the disease patterns in the field, noting that trees in areas exposed to the wind tended to be more badly infected than trees that were sheltered from the wind. Ward studied the coffee rust at both the large scale and the small scale simultaneously: "I carried microscopic and drawing apparatus everywhere, and, as always, carefully figured any objects of importance to the inquiry."[28] These studies helped guide his later inquiries. Before Ward had left for Ceylon, he met with Daniel Morris in London. After Morris had left Ceylon, planters realized that the chemical treatments he had advocated did not effectively contain or control the rust, so Ward was under tremendous pressure to find a solution.

Ward's experimental work focused on sorting out the life cycle of the fungus. Morris and others had argued that the disease partly manifested itself in a threadlike mycelium that hung off the leaves. Ward, however, doubted that this mycelium belonged to the coffee rust fungus. He found more than a dozen different kinds of spores on the coffee plants. "Obviously," he noted, "it will be necessary to trace the origin and destination of each form until we are either sure it is foreign or it is connected with 'leaf disease.' . . . Until we are quite satisfied as to which of the filaments, if any, belongs to the *Hemileia*, we are in the dark respecting one phase of its life-history, and until we know its life history, remedial measures are premature." He had attempted to infect coffee plants with the spores and the fungi he had collected, without much success.[29]

Ward had also begun experiments and analyses to show how the fungus harmed the coffee tree. With this experiment, he sought to refute those who argued that the fungus was a symptom of the disease, rather than its cause. Using spores he had collected in the field, he successfully infected a healthy coffee plant that had been kept isolated in a Wardian case. His microscopic studies of the leaves showed how the mycelium of the fungus fed off the cells in the coffee tree—effectively obliging the coffee leaf to feed the fungus as well as the plant. The leaves were essential for providing food for the branches and the fruit of the coffee tree; the action of the fungus, coupled with the premature leaf fall it provoked, starved the fruit and the branches of vital nutrition. Deprived of food, the production of fruit fell off, and in serious cases some of the branches died: "Exhausted on the one hand by crop, and on the other by fungus, no wonder the leaves become rapidly poor in food materials, turn yellow, and fall all along the branch up to the terminal bud: whatever sap then flows upwards will be used by the crop, and never reach the bud, which consequently dies." Ward said little about the practical consequences of these observations for planters, except to suggest that manuring would likely be of little help over the longer term, since "the continual renewal of leaf calls forth great energy on the part of the tree, and it no doubt follows as a direct consequence of this that the crop-producing capabilities of the tree are for a time lessened."[30]

Ward continued his work through 1881, and in September published his third and final report on the coffee rust. The report reviewed and summarized all the previous research. In it, Ward reconstructed *H. vastatrix*'s entire life cycle, reviewing both his experiments in the laboratory and in the field. Henry Trimen's introduction observed that "it is not too much to say that as regards the structure, circumstances, and habits of *Hemileia* on the coffee leaf, we are now completely informed." Indeed, for all of the controversy and protest that the report was to generate, everyone seems to agree that Ward had brilliantly and conclusively established the fungus's life cycle.[31]

In the *Third Report*, Ward explicitly and implicitly engaged with previous, and in his view, erroneous interpretations of the disease. The report systematically discussed how all the previous theories of the origin and development of the coffee rust could be explained by the action of the fungus. Indeed, there were many problems to explain—most significantly the disease's seeming variability and randomness, but also how some coffee farms were badly infected, while others escaped scot-free, and the relation between disease and climate, good years and bad. As in his second report, Ward argued that the disease was caused by the *H. vastatrix* fungus alone. He reiterated that the fungus's main impact on the coffee tree was to cause premature defoliation, which in turn slowed the development of the branches and berries.

Ward's report enjoyed the force it did because of his meticulous experimentation and observation on all facets of the question. His studies built from the microscopic examination of the fungus on slides, to experiments in Wardian cases on the veranda of his bungalow, to field experiments and observations in the Royal Botanic Garden at Peradeniya, to field experiments and observations on working coffee farms across the island. For example, he infected several experimental plants with the fungus, and then followed the development of the disease over the next several weeks. At every stage, he also kept track of the weather, showing how the fungus—and therefore the disease—developed more rapidly in humid weather. From an experiment on two leaves, he could conclude that "what planters term an 'attack' of leaf disease, i.e., a sudden outburst of the 'rust,' results from the coming to maturity at or about the same time of a series of mycelia which have been formed from the successful sowing of a certain number of spores; since all were exposed to similar conditions we must look for the origin of the rust to the conditions previously present."[32]

On the geography of the disease, he demonstrated the critical importance of wind to the disease's dissemination—arguing that the wind helped spread the spores throughout the island. Ceylon was subject to two monsoons, and planters had long associated the outbreak of the disease with the arrival of the monsoons. By using specific coffee estates as examples, he showed how areas exposed to the wind—ridges and hills—generally suffered more from the disease than did sheltered areas, such as valleys. He argued that these areas were exposed to more spores blown in by the wind; they also tended to be areas where there was mist and fog, which allowed the spores to geminate.[33]

Ward was one of the first scientists to connect crop epidemic with agricultural practices. In his reports and in subsequent scientific papers, he developed an ecological critique of coffee cultivation in Ceylon. He argued that the coffee rust epidemic had not been caused by the introduction of a new pathogen—he contended that the *H. vastatrix* fungus had been present in the jungles of Ceylon for a long time. Rather, the epidemic had been caused by the introduction of new forms of cultivation in the nineteenth century. He had found that even small windbreaks on coffee farms—a line of tall trees, for example—could check the rust's spread. Ward noted that "it is a matter of regret that such immense unbroken areas of coffee exist without break of any kind, and one can trace the swaying backwards and forwards of the spore-laden winds in consequence."[34] Hooker, Thwaites, and Morris had made similar connections between large expanses of monoculture and the development of epidemic diseases. To that extent, Ward's observation was not particularly new. What *was* new was that Ward had *demonstrated* the link, and *explained* how it operated through the dissemination of the spores of *H. vastatrix*.

Ward also addressed the practical question of how planters could cure or control the disease. Planters had long believed that manure could be a cure for the disease. "Manure," concluded Ward, "can in no way be looked upon as the cause of the disease or a cure for it: its proper action was that of a food."[35] Ward concluded that it was difficult, if not impossible, to try to kill the fungus while it was in spore form, or after it had established itself in the coffee leaf. The only time the fungus was vulnerable was in the brief day or two after the spore had germinated but before it had penetrated the leaf. There were also further challenges: since the spores grew so quickly, it was necessary to find a chemical that could cover the entire plant (especially the undersides of the leaves) and that would be persistent, acting for days and weeks after a spraying. To make the task even more daunting, the reagent "must be soluble, and yet at the same when it is most required to be active is *when the rains are continuously dissolving it, and running off with it in solution to the ground*."[36]

Having reached his conclusions, Ward decided to return to Britain. He had established the life history of the fungus, and explained how the disease operated in the field. As Trimen noted, "Mr. Ward looks upon the present report as final in its character, and does not anticipate any discovery of practical value would result from further work at *Hemileia*."[37] Some coffee planters seemed to accept Ward's conclusion, but still sought some practical advice on coffee cultivation in general. Other planters were not so willing to accept that Ward specifically, or science generally, had spoken the last word on disease control. They even questioned some of his experimental methods. Commenting on Ward's research tactic of deliberately infecting the coffee plant, one planter wrote, "This *artificial* test was not a fair one. Planters don't want to know whether if artificial means are employed all coffee trees are equally liable to take in the fungus, and to suffer from the disease, but whether they are so or not, when subjected to the ordinary operations and conditions of nature."[38] Nonetheless, the planters' public and private writings in the following years showed that they accepted the basic premise of fungal pathogenicity; most had quietly abandoned their earlier models of disease. But this does not mean that they accepted the findings of Ward—or any other scientist—uncritically. As far as planters were concerned, scientists were not the only people who could produce new knowledge about coffee cultivation. "We have now," wrote the planter G. A. Talbot, "all that can be taught us by scientific men about *Hemileia*, and it is for practical planters, in working their coffee, to study the disease. I must say, I think there is a good deal to be found out yet."[39]

In his classic history of plant pathology, *The Advance of the Fungi*, E. C. Large argued that Ward's work had broader importance in plant pathology for two reasons. Ward was the first plant pathologist to propose the *preventive* chemical spraying of crops—to prevent the pathogen from establishing itself in the host to begin with. Ward was also the first plant pathologist to articulate clearly the role of landscape change, particularly the spread of monocultures, in producing crop epidemics. "In pointing to the exclusive cultivation of single crops over unbroken areas as one of the chief causes for the advance of the fungi," writes Large, "Ward did not put forward any panacea; he stated a fundamental truth."[40] Ward's insight, in turn, depended on the idea that crop diseases were ultimately caused by pathogens. In earlier models of disease, the environment was assumed to act on the host directly. In Ward's model, the environment acted by shaping the movement and dispersal of pathogens. It is at this point that the canonical stories of coffee rust end. Ward's research was not, however, the definitive word on the epidemic.

New Nature, New Knowledge: Rust Research, 1880–1930

In the decade after Ward's departure, Ceylon's coffee industry collapsed. Most coffee planters in Ceylon switched to tea, which remains the island's leading export crop to the present. Some coffee planters, however, decided to continue cultivating coffee elsewhere in the tropics. Unfortunately, it seems that as they fled the coffee rust epidemic, they wound up carrying the disease with them. The epidemic appeared in India as early as 1870, in Sumatra in 1876, and in Java the following year. By the 1930s, the epidemic had spread across the Indian Ocean basin and the Pacific. The fungus devastated coffee farms from Uganda and the Philippines; some areas were able to stave off the epidemic for a short while, but ultimately it appeared in every coffee region in the Old World. Despite the economic threat that this disease presented, public botanic gardens and agricultural research institutions in the tropics or in the imperial capitals responded slowly and halfheartedly. Perhaps they did so in part because Ward's reports suggested that most attempts at disease control would be futile.[41]

But while official scientific research may have been slow, European coffee planters in the tropics continued to carefully observe conditions in the coffee fields, and to conduct experiments where possible. They were quick to adopt new scientific and agricultural innovations from Europe. For example, British planters in India began to experiment with the Bordeaux mixture—a copper-based spray developed in the 1880s—and saw some success in controlling the coffee rust. But chemical sprays required expensive labor, technology, and supplies, so they were not really cost-effective. In 1880, the planters of Ceylon also founded a journal, the *Tropical Agriculturist*, which included important agricultural research articles, as well as reports from European planters of tropical crops across the tropics. In this sense, the *Tropical Agriculturist* was a tropical version of the *Gardener's Chronicle*, which published horticultural pieces from both professionals and amateurs.[42]

Coffee farmers continued to search for a rust-resistant coffee, in spite of Ward's conclusion that all coffee plants were susceptible. In the field, coffee farmers noted that different species and varieties of coffee succumbed to infestations of *H. vastatrix* at different rates. These could not be explained away simply in terms of environmental differences in the coffee farms. As Europeans explored and colonized equatorial Africa, however, they discovered new varieties and species of coffee, and introduced some of them to cultivation. The first of these was an entirely new species of coffee discovered in West Africa, popularly known as Liberian coffee (*C. liberica*). Liberian coffee plants were much larger and more productive than Arabica coffee plants, although they produced beans of an inferior cupping quality. Importantly, Liberian coffee plants also appeared to be resistant to the rust. Planters in Ceylon introduced Liberian coffee to the island in the early 1870s. The plant initially seemed to weather the onslaught of the rust much better than did Arabica, but after a few seasons its resistance seemed to break down. In the 1890s, Liberian coffee was introduced into infected Arabica zones in the Dutch East Indies and British Malaya. Once again, the plants initially withstood the epidemic better than Arabica, but lost their resistance after several seasons.[43] Similarly, coffee planters in India discovered that even different *varieties* of Arabica coffee showed differing degrees of resistance. While most of the Arabica crop in southern India was devastated by the rust epidemic, British coffee planters discovered a variety (Old Chik) that seemed to resist the rust epidemic. Significantly, the Arabica coffee cultivated in British India was of a slightly different genetic stock than that cultivated elsewhere in the Old World. Coffee plants had been introduced to the Western Ghats of southern India by Muslim pilgrims several centuries before the Europeans arrived. Some European planters obtained their planting stock from native coffee gardens. As with Liberian coffee, however, the Old Chik gradually lost its resistance to the rust. It was, in turn, replaced by the Coorg and then by the Kent varieties of Arabica, each of which dominated for a couple of decades before losing its resistance to the rust.[44]

Resistance was not always a transient property. In 1900, coffee planters in the Dutch East Indies introduced yet another newly discovered coffee from Africa, *C. canephora var. robusta*, now known as Robusta coffee. Like Liberian coffee, Robusta plants were vigorous and high yielding, but produced beans of an inferior taste. Unlike the Liberian coffee and the resistant Arabicas, Robusta coffee did *not* lose its resistance to the coffee rust. Scientists working at the coffee experiment station in the Dutch East Indies noticed this, and began selection programs to improve the quality and yield of Robusta. Between 1900 and 1930, planters across Africa, Asia, and the Pacific cultivated Robusta on a large scale, allowing for a partial resurrection of the old "Arabica graveyards." In 1905 the Old World produced only 5 percent of the world's coffee; by the middle of the twentieth century it produced more than 25 percent. The expansion of Robusta drove most of this recovery.[45]

By the early twentieth century, coffee planters had clearly established the fact of disease resistance, if not the theory behind it. Planters had demonstrated in the field that there *was* differential individual susceptibility, that inherent qualities of the host plant *did* help shape the course of the epidemic. In discovering this, however, planters also uncovered a new problem: resistance was sometimes—but not always—ephemeral. In their publications and in government reports, planters attributed the loss of resistance to changes in the host plants. There was little reason to doubt this observation. In the half century between 1880 and 1930, only a handful of scientists across the world were actively studying the problems of coffee cultivation. Pioneering coffee research was conducted in the Dutch East Indies, home to a cluster of public and private agricultural experiment stations, many devoted to particular tropical crops.

Arguably the leading coffee researcher in the early twentieth century was P. J. S. Cramer (1879–1952) at the Dutch Botanical Garden at Buitenzorg, Java. In the first decade of the 1900s Cramer suggested that the breakdown of resistance in Liberian coffee might have been caused by changes in the fungus. He observed that "vigorous" Liberian plants brought to infected regions quickly succumbed to the rust. Conversely, when seeds of coffee plants in infected regions were cultivated in disease-free regions, they produced healthy, vigorous plants. Based on this, Cramer argued that the apparent breakdown of resistance was caused by changes in the fungus, rather than the host plant. But Cramer did not pursue his speculation in any further detail.[46]

It was not until the early 1930s that another scientist turned his full attention to the problem of the rust. Wilson Mayne was the coffee scientific officer at an experiment station run by the United Planters' Association of South India. As he studied chemical control of rust, he also explored the apparent breakdown of resistance in Old Chik and Kent varieties of Arabica. He inoculated samples of Coorg and Kent coffee with different specimens collected in different areas. He discovered that some rusts attacked only the Coorg coffees, while others attacked both Coorg and Kent. He concluded, as Cramer had several decades before, that the breakdown of resistance in the coffee plant was, in fact, a change in the *fungus*. Mayne argued that this represented the emergence of a new strain (or "race") of the rust fungus that was physiologically specialized in attacking a particular kind of coffee. These changes were subtle: in microscopic studies, he could find no anatomical differences between the different strains of rust. The only way to distinguish between the races was by observing how different strains of coffee responded to an infection. Mayne identified two races of coffee rust in a paper published in *Nature* in 1932. Later in the 1930s, he identified two more races. Since then, scientists have identified several dozen more.[47]

By the mid-1930s, then, scientific and popular ideas of crop diseases—and coffee rust in particular—had changed significantly since the mid-nineteenth century. The newer models of disease did not fundamentally refute the pathogenic models articulated by de Bary, Ward, and others. In these newer accounts, the fungus still remained a central causal agent, and environmental factors did play a critical role in shaping disease patterns. But in the new model, both the fungus and the host plant were shown in a much more complex relationship, involving physiologically specialized races of pathogens attacking host plants that themselves has specialized resistance. These changing theories of disease, however, involved more than scientists and planters simply uncovering a preexisting nature waiting to be discovered. Rather, the increasingly complex theories and models of plant disease reflected *changes in nature itself*—largely anthropogenic changes. Changes in disease patterns reflected changes in nature. When planters introduced new varieties and species of coffee to the field, and cultivated them in specialized landscapes, they changed the genetics of resistance in local populations of coffee. Over many growing seasons, this created selection pressure on the populations of the rust fungus, which in some cases produced new rust races that could overcome the resistance of the host plant. In short, the scientific *theories* about the disease changed because the *disease itself* had changed, and the disease had changed because the field had changed.

Epilogue: The Science of Working Landscapes Since 1930

The interplay of scientists, farmers, and landscapes has, of course, continued to evolve since 1930. The coffee rust epidemic subsequently spread to West Africa, and then leaped across the Atlantic to reach the Americas some time around 1970. Over the next decade, the disease swept across the New World, although it did not cause as dramatic losses as it had in Ceylon and other parts of the Old World. New tools were introduced into the coffee landscapes on a large scale, such as the use of chemical fungicides. Similarly, new rust-resistant varieties of coffee were developed and tentatively introduced into production. In the 1970s and 1980s, many coffee farmers in the Americas "technified" their farms, introducing high-yield, dwarf hybrid coffees, eliminating shade trees, and using agricultural chemicals on a large scale. More recently, some researchers have been discussing the use of genetically modified coffee to address the rust epidemic and to deal with other problems. The scientific and technical changes that coffee cultivation has undergone since 1930 are not unique to coffee; corn and rice cultivation underwent similar transformations during the Green Revolution of the 1960s and 1970s.

The *power relations* that help shape these working landscapes have also changed. Across the twentieth century, institutions of one sort or another have played an increasingly important role in shaping agricultural landscapes, and also in shaping the relationships between scientists and farmers. In the banana industry, for example, large corporations such as the United Fruit Company shaped the landscapes, and also determined how banana planters would respond to the devastating crop epidemics that struck several times during the twentieth century. In Brazil, the Ford Corporation tried—and failed—to find a means of controlling the South American leaf blight that chronically plagued the Amazon's rubber plantations.[48] Multilateral organizations such as the Food and Agriculture Organization of the United Nations, and bilateral development organizations such as the United States Agency for International Development, have also done much to shape agricultural landscapes in the twentieth century.

USAID and the Inter-American Development Bank, for example, actively sponsored the technification of coffee production in the 1980s and 1990s. Diseases such as the coffee rust continue to plague many tropical crops, and will likely continue to do so for the foreseeable future in spite of the best efforts of farmers and scientists. These diseases will limit the power both of large institutions and of small farmers alike to shape their working landscapes as they please.

Notes

1. Not all historians—or even all field scientists—would agree with this claim. Philip Pauly is one historian whose work spans both the history of biology and the environmental history of working landscapes. See his reflection "Is Environmental History a Subfield of Garden History?" *Environmental History* 10 (2005): 70–71.
2. Richard Grove, *Green Imperialism: Colonial Expansion, Tropical Island Edens, and the Origins of Environmentalism, 1600–1860* (Cambridge: Cambridge University Press, 1995); and Margaret Rossiter, *The Emergence of Agricultural Science: Justus Liebig and the Americans, 1840–1880* (New Haven: Yale University Press, 1985).
3. James C. Scott, *Seeing Like a State: How Certain Schemes to Improve the Human Condition Have Failed* (New Haven: Yale University Press, 1998).
4. Christy Campbell, *Phylloxera: How Wine Was Saved for the World* (New York: HarperCollins, 2004).
5. Stuart McCook, "Global Rust Belt: *Hemileia vastatrix* and the Ecological Integration of World Coffee Production Since 1850," *Journal of Global History* 2 (2006): 177–195.
6. James L. A. Webb Jr., *Tropical Pioneers: Human Agency and Ecological Change in the Highlands of Sri Lanka, 1800–1900* (Athens: Ohio University Press, 2002), chap. 3; and P. J. Laborie, *The Coffee Planter of Saint Domingo* (London: Cadell and Davies, 1797).
7. G. C. Ainsworth, *Introduction to the History of Mycology* (Cambridge: Cambridge University Press, 1976), chap. 6; Ainsworth, *Introduction to the History of Plant Pathology* (Cambridge: Cambridge University Press), 31–38; and E. C. Large, *Advance of the Fungi* (London: Jonathan Cape, 1940), 63–65, 96–103, 131–136. For an older overview, see Herbert Hice Whetzel, *An Outline of the History of Phytopathology* (Philadelphia: W. B. Saunders, 1918).
8. Miles Joseph Berkeley, "Observations: Botanical and Physiological, on the Potato Murrain," *Journal of the Horticultural Society of London* (January 1846), quoted in Ainsworth, *Introduction to the History of Mycology*, 155.
9. Broome to Berkeley, 1 September 1869, Berkeley Papers, Natural History Museum, London.
10. M. J. Berkeley, *Gardeners' Chronicle*, November 6, 1869, 1157.
11. Ibid.
12. On the economics of coffee in the early 1870s, see Alastair Mackenzie Ferguson and John Ferguson, *The Planting Directory for India and Ceylon* (Colombo, 1878), 35.
13. G. H. K. Thwaites, "Scientific Committee," *Gardeners' Chronicle*, May 4, 1872, 609.
14. G. H. K. Thwaites, "Ceylon Coffee Fungus," *Gardener's Chronicle*, 1874, 726.
15. D. Morris, "Coffee-Leaf Disease of Ceylon and Southern India," *Nature* 20 (1879): 558.
16. Thwaites, "Ceylon Coffee Fungus," 726.

17. "Government and Leaf Disease," *Ceylon Observer*, February 19, 1879, in Kew Papers, Royal Botanic Gardens, Kew (hereafter cited as Kew Papers), Misc. Rept. Ceylon Coffee Diseases, fol. 60.
18. Richard Drayton, *Nature's Government: Science, Imperial Britain, and the "Improvement" of the World* (New Haven: Yale University Press, 2000), 244–248, 252.
19. G. H. Thwaites to the director, Royal Botanical Garden [1879], Kew Papers, Ceylon Coffee Diseases, fol. 164–165.
20. Morris, "Coffee-Leaf Disease," 559; and Daniel Morris, "Notes on the Structure and Habit of *Hemileia vastatrix*, the Coffee-Leaf Disease of Ceylon and Southern India," *Journal of the Linnaean Society—Botany* 17 (1879): 512.
21. Webb, *Tropical Pioneers*, 112.
22. Quoted in "Coffee Leaf Disease," *Gardeners' Chronicle*, May 3, 1879, 564.
23. "Coffee-Leaf disease," *Gardeners' Chronicle*, August 23, 1879, 240.
24. Quoted in "Coffee Leaf Disease," *Gardeners' Chronicle*, May 3, 1879, 564.
25. Quoted in Drayton, *Nature's Government*, 250.
26. Frederick Lewis, *Sixty-four Years in Ceylon: Reminiscences of Life and Adventure* (Colombo, Ceylon: Colombo Apothecaries, 1926), 105.
27. On Ward's appointment, see William Thiselton-Dyer to Secretary of State, 9 October 1879, Kew Papers, fol. 131–132. One of the first biographical sketches of Ward was written by Thiselton-Dyer, "Harry Marshall Ward, 1854–1906," in *Makers of British Botany: A Collection of Biographies by Living Botanists*, ed. Francis Wall Oliver (Cambridge: Cambridge University Press, 1913), 261–279. This brief essay remains one of the most valuable sources for this subject. The first book-length study of Ward and his work is Peter Ayers, *Harry Marshall Ward and the Fungal Thread of Death* (St. Paul, Minn.: American Phytopathological Society, 2005).
28. Henry Marshall Ward, *Preliminary Report on the Enquiry into the Coffee-Leaf Disease in Supplement to the Ceylon Observer* (Colombo, Ceylon: A. M. & J. Ferguson, 1880), Kew Papers, Ceylon Coffee Disease, fol. 177.
29. Ibid, fol. 178.
30. H. Marshall Ward, *Coffee Leaf Disease: Second Report*, Ceylon Sessional Paper 50 (Colombo, Ceylon: Government Printer, 1880), 6, 8.
31. Trimen, introduction to H. Marshall Ward, *Coffee Leaf Disease: Third Report*, Ceylon. Sessional Paper 17 (Colombo, Ceylon: Government Printer, 1881), 1.
32. Ward, *Third Report*, 8.
33. Ibid., 13.
34. Ibid., 15.
35. Ibid., 4.
36. Ibid., 18–19.
37. Trimen introduction, in ibid., 4.
38. "The Cryptogamist and the Leaf Disease," *Tropical Agriculturist*, February 1, 1882, 730.
39. G. A. Talbot, "Mr. Marshall Ward's Report on Leaf-Disease," *Tropical Agriculturist*, January 2, 1882, 626–627.
40. E. C. Large, *The Advance of the Fungi* (London: Jonathan Cape, 1940), 205.
41. William Gervase Clarence-Smith, "The Coffee Crisis in Asia, Africa, and the Pacific, 1870–1914," in *The Global Coffee Economy in Africa, Asia, and Latin America,*

1500–1989, ed. William Gervase Clarence-Smith and Steven Topik (Cambridge: Cambridge University Press, 2003), 100–119; and McCook, "Global Rust Belt."

42. McCook, "Global Rust Belt," 182–187.
43. On Liberian coffee, see G. A. Cruewell, *Liberian Coffee in Ceylon: The History of the Introduction and Progress of the Cultivation up to April 1878* (Colombo, Ceylon: A. M. & J. Ferguson, 1878); Frederick L. Wellman, *Coffee: Botany, Cultivation, and Utilization* (London: Leonard Hill, 1961), 76; and P. J. S. Cramer, *A Review of Literature of Coffee Research in Indonesia* (Turrialba, Costa Rica: Inter-American Institute of Agricultural Sciences, 1957), 43–45, 105–112. See also Francis B. Thurber, *Coffee: From Plantation to Cup*, 14th ed. (New York: American Grocer Publishing Organization, 1887), 107–116.
44. This sequence is summarized in Frederick L. Wellman, *Tropical American Plant Disease* (Metuchen, N.J.: Scarecrow Press, 1972), 494–495.
45. For a recent account of *C. canephora* and rust resistance, see Albertus B. Eskes, "Resistance," in *Coffee Rust: Epidemiology, Resistance, and Management*, ed. Ajjamada C. Kushalappa and Albertus B. Eskes (Boca Raton, Fla.: CRC Press, 1989), 171–292. See also Cramer, *Coffee Research in Indonesia*, 45–46, 119–140.
46. P. J. S. Cramer, "L'influence de l'*Hemileia vastatrix* sur la culture du café au Java, Partie I," *L'Agronomie Tropicale* (Brussels) 2 (1910): 345–346.
47. W. W. Mayne, "Physiological Specialization of *Hemileia vastatrix* B. & Br.," *Nature* 129 (1932): 510; and Mayne, "Recent Work on Coffee Leaf Disease," *Planters Chronicle* 27 (1932): 253–257. For an overview of resistance, see Eskes, "Resistance," 211–215. The concept of physiological races was, by that point, not completely new to plant pathology. Researchers working on the wheat rust had first identified physiological races in the mid-1890s. For a succinct discussion of physiologic races of plant pathogens, see Ainsworth, *Introduction to the History of Plant Pathology*, 48–51.
48. John Soluri, *Banana Cultures: Agriculture, Consumption, and Environmental Change in Honduras and the United States* (Austin: University of Texas Press, 2005); and Warren Dean, *Brazil and the Struggle for Rubber* (Cambridge: Cambridge University Press, 1987).

Five

Jeremy Vetter

Rocky Mountain High Science

Teaching, Research, and Nature at Field Stations

Are there field sites where scientists can produce knowledge about global environments in one place? Today, field scientists often go into the mountains to study the effects of global climate change. It is on mountaintop biological islands that the devastating effects of climatic warming are often felt the earliest, since plants and animals adapted to such places usually have nowhere to go. Even when the effects are less dramatic than (local or global) species extinction, mountain environments are often considered especially sensitive to changes in the global climate.[1] The use of mountain stations for long-term climate change research reveals how particular local field sites can illuminate critical global environmental problems. Yet using mountains as places for fieldwork on global environments is far from new: the promoters of Rocky Mountain field stations promulgated such a vision in the early twentieth century.

When early twentieth-century field scientists established mountain stations, however, they promoted them less as places of greater local environmental sensitivity and more as places of greater global environmental representation. Such a representation of diverse environments was achieved not through the coordination of fieldwork in many different places distributed around the world, but rather by making local field sites produce high-level science on their own. Such field sites have been strategically located to provide access to *multiple* environments, elevating the epistemic stakes and enabling field researchers to leverage work of broad purported significance out of sites that were undeniably local or, at best, regional. In a sense, they were also attempts to bridge what one might call the "epistemic rift" between the residential knowledge gained through direct, ongoing experience of a place and the cosmopolitan, globalizing knowledge of science.[2]

In the U.S. Rocky Mountains, for example, which is the regional focus of this essay, researchers set up field stations, especially in the life sciences, to produce a kind of high science that was simultaneously local in practice, regional in the political and economic forces that shaped it, and transcontinental—indeed, almost global—in its epistemic aspirations. One such field site was the University of Colorado's Mountain Laboratory, located at Tolland, a small village in a high mountain valley of the Colorado Front Range. Founded in 1909, the Mountain Laboratory was directed for ten years in that location by the university's professor of biology, Francis Ramaley (1870–1942). The advantages of place were central to the appeal of Tolland as a site for serious field research. "Almost at his door," Ramaley proclaimed, "the student has as wide diversity in climate as could be found in a trip from Illinois to Ellesmere."[3] Few other places on earth offered such an array of contrasting places within an accessible radius. It was on such a foundation that Ramaley and other Rocky Mountain field station proponents articulated the wider significance of their activities.

Field stations in the Rocky Mountains undoubtedly have produced high science in the literal sense of physical elevation above sea level. My interpretation in this essay, however, revolves around a second meaning as well: the field scientists' ambition to construct high science in a more fundamental, epistemic sense. It might seem strange to look for high science at biological field stations on the periphery of the American scientific landscape, and moreover, at institutions that offered a messy combination of teaching, research, and summer vacationing. In the grand sweep of the history of science, to examine field stations seems to indicate a dramatic descent from the metropolitan realms of high theory and cutting-edge laboratory experimentation. Yet historians can understand the Rocky Mountain stations as places where researchers attempted to legitimate a high science of the field. Indeed, the scientific rigors intended for Rocky Mountain field stations sometimes even meant naming them with the word "Laboratory"—examples included not only the University of Colorado's Mountain Laboratory, but also Frederic Clements's Alpine Laboratory at Minnehaha, Colorado, and the Rocky Mountain Biological Laboratory at Gothic, Colorado—thus effectively designating them as hybrid "labscapes."[4] Nevertheless, the Rocky Mountain field stations remained among the most field-like of all labscapes because of their deep, intimate relationship with surrounding environments.

The lofty status of Rocky Mountain field station science is most clearly visible from a regional point of view. From a European or eastern U.S. metropolitan perspective, the field station was in a peripheral location, fit primarily for modest, low-level fact gathering in anticipation of metropolitan theorizing. Yet from the perspective of the interior American West, the field stations analyzed in this essay were strikingly different from the other kinds of stations that predominated elsewhere within this region, most of which were branch agricultural experiment stations, along with a few counterparts in forestry, mining, and other economic sectors. These more typical stations were tied directly to practical development concerns, and the height of their scientific ambitions was severely restricted by their primary focus on adapting knowledge to local conditions (think of variety testing at branch agricultural stations).[5] Thus, viewed from such a perspective, the ambition and design of the Rocky Mountain field stations to produce disciplinary knowledge for national scientific journals were striking.

But how far did the Rocky Mountain field stations succeed in their high epistemic aspirations? Did the access they provided to a diverse range of environmental conditions enable them to produce knowledge that would be applicable on a large geographical scale? Did they help field scientists to know global environments? These questions can best be answered through a close investigation of how such field sites operated: how they integrated teaching and research, and the practices that enabled and mediated scientific experiences of nature in settings usually better known for their recreational opportunities.

The Field Station Phenomenon

In exploring how the Rocky Mountain field stations aspired to a form of high science on the periphery, it is essential to begin by observing two other noteworthy features of their funding and viability. First, even high science relied on a practical justification for its support. As for many museums and state surveys, the favored justification was education. Even within the broader category of education, however, the Rocky Mountain field station aimed high: its students would be largely college students and graduate researchers, not the schoolchildren and general public that the museums and state surveys focused so much on.[6] In this sense, even within the universe of institutions with educational missions, the field station was envisioned as a crucible for high science pedagogy. Just as graduate education for research degrees in the United States was built on the base of the elective system for undergraduate course teaching—what Kohler memorably dubbed the "Ph.D. machine"—so too did field stations build their serious, advanced research on a foundation of summer school for their viability.[7] Second, teaching was undeniably a practical context of sorts, but it seems evident that station leaders deemed it more compatible with prestigious research than work at other field stations, with their abundant ties to agriculture, mining, forestry, and other profit-making domains.

As one of the foremost American architects of the pure science ideal in the late nineteenth century proclaimed, "The duty of a professor is to advance his science, and to set an example of pure and true devotion to it which shall demonstrate to his students and the world that there is something high and noble worth living for."[8] Explicitly contrasted with the pursuit of applied knowledge for making money, higher education was widely maintained as a force for moral improvement and uplift. Building a field station on an educational base more neatly preserved a rhetorical distinction between science and capitalism, thus elevating the work done there to higher status.

The earliest field station type was the marine biological station, epitomized by Woods Hole, Massachusetts, a great "summer resort" for serious scientific research work.[9] As lake and mountain stations became appealing as alternatives to seaside biological stations, they proliferated around the turn of the twentieth century.[10] (Prestigious, high-science research also flourished at the Desert Botanical Laboratory, founded near Tucson, Arizona, in 1903 and funded by the Carnegie Institution of Washington.)[11] Field stations attempted to extend the rigor of the laboratory into the field.[12] Moreover, field stations often came to represent entire regions for practitioners of the newly emerging discipline of ecology.

The Rocky Mountain region was by no means a new field research frontier at the end of the nineteenth century. Numerous explorations and surveys had long studied the region's geology, plants, animals, and other natural features.[13] In conjunction with survey work, temporary triangulation stations were established on mountaintops and other advantageous locations, and these stations had occasionally produced special scientific observations. For example, a July 1894 electrical storm on the summit of Mount Elbert—Colorado's highest peak—experienced by one member of the U.S. Coast and Geodetic Survey even reached the pages of *Science*.[14] Even before that, Colorado's Pikes Peak had served as the site for a long-term meteorological observatory operated by the U.S. Army Signal Corps.[15] With few exceptions, however, fieldwork in the nineteenth-century Rocky Mountains was based on temporary occupancy of high mountain locations for short-term research goals.

By the early twentieth century, the region became a fertile domain for the development of more permanent field stations, especially though not exclusively in the life sciences. The first two known full-fledged Rocky Mountain field stations were a New Mexico station near Las Vegas (relocated there in 1898 from the southern part of the state) and a new hybrid, lake-and-mountain station in Montana in 1899, established at Flathead Lake just west of the Continental Divide and south of Glacier National Park.[16] That same year, Frederic Clements and other field researchers from the University of Nebraska first

Table 5.1
Field Stations in the Rocky Mountains

Date	Name	Location	Sponsoring Institution
1899	New Mexico Biological Station	Las Vegas, N.Mex.	New Mexico Normal Univ.
1899	U. of Montana Biological Station	Flathead Lake, Mont.	University of Montana
1899	Alpine Laboratory	Pikes Peak, Colo.	U. Nebraska/ Minnesota/CIW
1906	Camp Colorado	Ute Pass, Colo.	Colorado College
1909	Mountain Laboratory	Tolland, Colo.	University of Colorado
1914	Mountain Laboratory	Silver Lake, Utah	University of Utah
1922	Science Lodge	Nederland, Colo.	University of Colorado
1922	Rocky Mtn. Biological Station	Almont, Colo.	Western State Coll. of Colo.
1922	Rocky Mtn. Summer School	Palmer Lake, Colo.	McPherson College (Kan.)
1923	Summer Science Camp	Medicine Bow, Wyo.	University of Wyoming
1928	U. of Denver Biological Station	Estes Park, Colo.	University of Denver
1928	Rocky Mtn. Biological Laboratory	Gothic, Colo.	Independent
1929	Camp Davis	Hoback Jct., Wyo.	University of Michigan
1935	Geology Camp	State Bridge, Colo.	University of Michigan
1936	Mount Evans Laboratory	Mount Evans, Colo.	University of Denver/MIT
n.d.	Field School of Biology	Jemez Springs, N.Mex.	Univ. of New Mexico

Source: Homer A. Jack, "Biological Field Stations of the World," *Chronica Botanica* 9 (1945): 63, 70–71, 73, with additions from other sources.

used Minnehaha, Colorado, in the Pikes Peak region, as a base for their ecological research work. In the next few decades, several more stations proliferated in high mountain locations across the Rockies. Most of these were sponsored by colleges or universities within or near the region itself (see table 5.1).

Figure 5.1

Francis Ramaley (*center*), and two colleagues sitting on the steps of Mountain Laboratory in Tolland, Colorado, 1909.
Archives, University of Colorado, at Boulder Libraries, Francis Ramaley Papers, Box 4, Envelope 8.

These field stations varied considerably in many respects, including their longevity, their size, and their balance of teaching and research. As noted above, I am focusing especially on one particular station in the Colorado Rockies: the Mountain Laboratory of the University of Colorado, located in Tolland (see figure 5.1). Other stations in the region serve to bolster my arguments at various points by similarity or contrast. An especially noteworthy comparative case is the Alpine Laboratory operated by pioneering American ecologist Frederic Clements—at different times associated with the University of Nebraska, the University of Minnesota, and the Carnegie Institution of Washington (CIW)—near Pikes Peak, Colorado. Due to its widely recognized role in the development of Clementsian ecology, and funded by the CIW as an independent research facility during much of its existence, the Alpine Lab was arguably the most important in the region.[17] But I am focusing here on the Mountain Lab, which is more typical of the Rocky Mountain field stations, depending as it did on university support as an outpost of summer course teaching. If Clements was unusual for his strong research funding and CIW patronage, and if many other station leaders were often too obscure to appear much at all in the national disciplinary journals, then perhaps the Mountain Laboratory under Ramaley serves as a suitable intermediate case.

Powers of Place

What was so appealing about the Rocky Mountains as a place to locate summer field stations? Most important from the point of view of both research and pedagogy was the availability of diverse environmental resources along the steep slopes from the foothills to the alpine tundra. "From the Great Plains grasslands," claimed a notice for Clements's Alpine Lab, "the series [of formational zones] runs from valley woodland at 5,800 feet to mesa, chaparral, foothill woodland, pine forest, aspen woodland and spruce forest to alpine meadow, rock field and bog at 11,000–14,000 feet in a distance of 7 miles."[18] Likewise, Ramaley's Mountain Lab was close to subalpine lakes, creeks, "a broad meadow," the mountains themselves, a river valley, and "a succession of forest-clad hills."[19] The Flathead Lake site for the University of Montana biological station similarly offered "a combination of lakes, rivers, mountains, forests, at elevations from 3,000 to 10,000 feet, [which] one will find in few places in America."[20] The Colorado Front Range sites were especially favored as "ideal places for field experiment" because "their slopes arise directly from the plain," so that they "contain in miniature the habitats and formations found between latitude 40 [degrees] and the arctic circle."[21] Station directors, in a line of reasoning that traces back to the fieldwork of Alexander von Humboldt, if not earlier, touted the tremendous altitudinal variation as equivalent to vast latitudinal traverses of the Earth.

Station directors emphasized both the research and teaching benefits of such a well-placed location amid so many diverse environments. These advantages were expressed not only in the ordinary language terms mentioned above but also in the more formalized terminology of ecological life zones. Ramaley, in fact, used the zones he identified around the Mountain Laboratory as the basis for articulating a set of distinctive vegetation zones found throughout the Rocky Mountains in North America. As Ramaley and a colleague noted in 1909: "Tolland is in the center of a most interesting botanical field, the limits of which on the one side are the foothill districts, on the other the Alpine heights, with the intervening montane and sub-alpine regions, all within easy reach."[22] Likewise, the Mountain Laboratory's nearby successor, the University of Colorado's Science Lodge, was later described as "offer[ing] the best of opportunities for the study of plant and animal distribution in relation to climate, because so many different climates exist within a distance of a few miles." No fewer than "five life zones" were accessible in a twenty-five-mile radius: "From the Upper Sonoran of the plains east of Boulder . . . through the Transition, Canadian, and Hudsonian, to the treeless Alpine of the Peaks of the Continental Divide," all accessible "in only a matter of a few hours."[23]

Using elevation change on mountain slopes as a proxy for a wide spectrum of global environments has had a long history in field practice. By the early twentieth century, mountain slopes were being adopted as places for field research by scientists concerned with studying the effects of environmental variation on the living world. Within the American West, for example, the diversity of environments available along gradients from mountaintop to plain attracted serious attention from researchers such as Clements at his Alpine Lab and his onetime collaborator Harvey M. Hall further west in California. Both Clements and Hall established chains of field sites—experimental gardens at various elevation levels—as part of their systematic work in cutting-edge, "experimental taxonomy" starting in the late 1910s.[24] Though such elaborate arrangements may have proven difficult to manage, they revealed the larger ambitions of some especially innovative and systematic American biologists to leverage the range of diverse environmental conditions available in mountain zones. Clements, Hall, and other ecologists used such mountainside experimental setups in the field to produce major works within their own programs of original research.

The theme of access to diverse environments, often with formal ecological zone names, was picked up by stations elsewhere in the Rockies. At the Rocky Mountain Biological Laboratory in Gothic, Colorado (near Crested Butte, north of Gunnison), leaders touted the elevation change from eight thousand to fourteen thousand feet, which generated a "great diversity in fauna and flora and offers examples of many different ecological communities," noting elsewhere in a draft promotional announcement that the "range of environmental variation is so great that one passes in two hours of time through several floral and faunal zones."[25] Likewise, the University of Wyoming's Summer Science Camp offered a "diversity of life zones, including plains, foothills, mountains, sub-Alpine and Alpine." Activities at the Summer Science Camp involved both teaching and research. "Within a radius of ten miles," proclaimed one announcement, "the region ranges in elevation from 7,000 feet to 12,000 feet." Most of this natural diversity was "easily accessible by numerous roads and trails and offers an excellent opportunity for the study of vegetation and animal types of the various life zones." Not only were varied plants and animals accessible on land, but so too were water resources, including "more than a hundred lakes, many hundreds of ponds, and numerous mountain streams within the ten mile radius," and even geological resources, including "a great diversity of rocks, which aggregate more than 50,000 feet in thickness." Despite the Camp's suspiciously low-status formal name, station promoters also dubbed it a "Laboratory," as with other field stations, to display the station's serious research aspirations. "No region could be better situated for a mountain Biological Laboratory," announced the organizers.[26]

Yet this rhetoric was also accompanied by an alternative legitimation strategy based on the special features of the mountain environment rather than special access to multiple environments. Thus, even as Ramaley pointed out the many environmental regions accessible from the Mountain Laboratory, he insisted that the station was "in no way a rival of the many excellent seaside and lakeside laboratories in various parts of the world." Notwithstanding the existence of Clements's more-exclusive Alpine Laboratory, Ramaley contended that the Mountain Laboratory "occupied a unique position in affording opportunity for the study of the rich flora and fauna of a mountain district." Pedagogically, it "place[d] the student where he must appreciate the relationship which organisms bear to their environment."[27] Another commonly touted feature was the possibility of new, undiscovered species nearby, although such a strategy would presumably backfire if the site became too closely associated with old-fashioned natural history practices rather than cutting-edge ecology. For example, promoters of the Rocky Mountain Biological Lab at Gothic described the entire area—mostly public land—as "virgin soil for the research botanist and zoologist . . . an area larger than the state of Connecticut that has never been investigated."[28] In other words, the Rocky Mountain field station not only could facilitate access to diverse environments but also could allow researchers and students to investigate the special ecology and biological species of the mountains.

The place-based advantages of the Rocky Mountain station as a site for field research and pedagogy were the most central benefits, but they were not the only ones. Secondarily, but not inconsequentially, the mountain field stations also offered an appealing locale for summer living. As environments associated with middle-class outdoor recreation—much like the seaside biological stations—mountain sites attracted students and professors alike.[29] The Mountain Laboratory in Tolland, Colorado, for example, offered "a comfortably cool climate in which to work during the heated months of the year."[30] Elsewhere, Ramaley and a colleague expanded at length on this basic formula, revealing a deeper reasoning linking environment with mental and bodily health in summer, aiming their appeal especially to schoolteachers and amateurs:

> It is not always possible to study in summer time without feeling the brain fag which comes with exertion in hot weather. The cool breezes of even our

Figure 5.2

Park Lake with town of Tolland in background, 1909.
Archives, University of Colorado at Boulder Libraries, Francis Ramaley Papers, Box 3, Envelope 41.

> most northern cities may be lacking in July and August, and the student who spends six weeks at a summer school may wonder whether he has really gained enough in mind to compensate for the wear and tear of body. For those who would study the outdoor sciences the question is a vital one. Only the most enthusiastic devotee of nature can enjoy long field trips in the hot, moist weather of mid-summer; yet this is the only time when the teacher or the amateur botanist can get the introduction which may be needed for a better grasp of the subject.[31]

As this passage indicates, summer fieldwork was a human residential experience that produced knowledge with recreational use value, in addition to offering chances for researchers to produce disciplinary knowledge that had a larger exchange value within cosmopolitan science.

This general pitch was supplemented with a barrage of specific information advertising the Tolland area (see figure 5.2) as a summer vacationing resort: "The town lies nestled at the base of a high mountain in a small valley, through which flows a trout-filled stream named South Boulder Creek." The June and

July average temperature at Tolland was "15 degrees lower than that of New York City . . . with plenty of warm, sunshiny days and clear, cool nights," thus providing "almost ideal conditions . . . for study, both indoors and out."[32] Similarly, Western State College advertised its Rocky Mountain Biological Station's location at tiny Almont, the "heart of the best fishing on the Taylor and East rivers," along with scenic drives and close proximity to beautiful Taylor Park.[33] The Rocky Mountain field station was therefore more than simply a site for producing knowledge; it was equally a place for enjoying long-term experiences of nature in residence that could not be enjoyed in the urban environments in which participants usually lived.

Not only were the summer conditions at Tolland and other sites in the Rocky Mountains contrasted with those in eastern cities, but they were compared favorably with those at better-known resort locations as well. "Climate cool & invigorating resembling that of Canadian lake resorts but drier," read one brochure for the Mountain Lab.[34] Moreover, the *Plant World* feature article advertising the Tolland station unashamedly acknowledged that they anticipated many students would come "partly for the outing and will not care to work all day." Moreover, they would have the chance to "ascend James' Peak and other mountains" of the Colorado Front Range, perhaps even hike to the "Arapahoe Glacier, 12 miles distant, the most southern glacier in the Rocky Mountain region."[35] Such excursions could have pedagogical, or even original research, purposes in many cases, but the station directors appealing for summer students did not hold back in promoting such activities as enjoyable and stimulating recreation, quite distinct from their investigative value.

Other Rocky Mountain field stations offered similar justifications. The Summer Science Camp of the University of Wyoming was located thirty-five miles west of Laramie on twenty-five acres carved out of U.S. Forest Service land, with a "delightful summer climate" at an elevation of 9,500 feet.[36] The Science Lodge, successor to the Mountain Laboratory from the 1920s onward, was located "on the flank of Mount Niwot, just below timberline, assur[ing] a bracingly cool summer climate." The station was "surrounded by rugged peaks, snowbanks, and a living glacier above timberline; by lakes, and forests threaded with mountain streams, trails, and roads below timberline." The Science Lodge offered both the "comforts of home" and the "freedom of camp life." Recreation was explicitly mentioned, including indoor and outdoor games, "hiking, lectures, and occasional cooperative trips to Boulder and other towns for movies, plays, or dances." Since by this time the camp could rely on automobile access, the Science Lodge summer program also promised an "expedition" to see plays at the "famous old Opera House at Central City, Colorado," many tens of miles distant.[37] Such attractions were generally touted more explicitly to prospective summer students rather than researchers, but since the former were often essential to the viability of the station, it was a crucial part of the package—and we can only assume that such activities were not wholly lacking in enticement for researchers too.

In the era before the automobile could easily transport vacationing students to the Central City Opera House, their accessibility from the outside world by rail was also crucial to the success of biological stations in the Rocky Mountains. The founder of the University of Montana Biological Station described it as "one of the most convenient places to reach from the Great Northern railroad on the north, and the Northern Pacific on the south," even though it was not itself on a railroad line.[38] In Colorado, both the Mountain Laboratory and the Alpine Laboratory were directly located on railroad lines. These rail access corridors were important not just to bring people and supplies in, but also to transport them more quickly across the wide variation of environments that formed such a crucial part of the Rocky Mountain station's justification. This combination of tremendous environmental diversity, discussed above, and a functional local transit system, featuring two concentric zones of walking and train riding, was neatly encapsulated by Ramaley and his colleague in a passage that captures the place-based system for extending the local field site:

> The hot, arid plains or the alpine heights are within a few hours ride by train. All of the life zones are so compressed that one may walk from Tolland in the montane zone down the valley for three miles to a typical foothill district, or up the valley for the same distance into the sub-alpine forest. A ride of one hour by train takes the student to the plains, while the hour and a half is sufficient to reach the top of the snowy range, or Continental Divide. Either of these excursions takes the traveler to a region as different from that of Tolland as Maine is different from South Carolina; while in passing the whole distance from cactus plants to alpine tundra, there are as great changes in the plants as would been seen in a journey all the way from Louisiana to Hudson's Bay.[39]

Could a seaside resort on the coast of South Carolina, Louisiana, Maine, or Hudson's Bay compete with that? Despite the occasional demurrals of some Rocky Mountain field station directors, it is hard to believe that such comments were not at least in part directed at proving the superiority of their own field sites for serious teaching and research compared to better-established seaside and lakeside stations. Moreover, they show how a residential mountain field station might provide a base for travel to all these diverse environments within the compass of everyday experience.

The Role of Teaching

So far I have largely evaded an issue that has long been central to the interpretation of American biological field stations of all kinds: the role of teaching. From the earliest marine stations of the East Coast, U.S. field stations were usually tied to teaching activities—at first, secondary-school teachers, and later including current college and university students on a variety of occupational tracks.[40] Indeed, fully three out of four American field biology stations were partly or completely oriented toward teaching.[41] The stations of the Rocky Mountains fit the pattern: most had a strong teaching component, and some were devoted entirely toward teaching. Since almost all were affiliated with particular colleges or universities, it was teaching—offering summer school courses off-site—that usually provided the primary rationale for institutional support. (As for the marine stations, providing continuing education to schoolteachers could also be important.) Rocky Mountain stations differed in how they combined teaching with research, but nearly all had a teaching component of some kind. It would be erroneous to see the aims of "teaching" and "research" as ineluctably competing for dominance. On the contrary, station proponents often argued for their symbiosis, and the ways in which they were integrated—bridging the epistemic rift between broad everyday experience and specialized research—show a range of creative options for combining the two allegedly distinct activities.

At the Mountain Laboratory, the University of Colorado's regents authorized the establishment of a field station "in connection with the summer session" of 1909, renting a "suitable building" and holding some classes there. These summer courses were the foundation of station activity throughout the Laboratory's existence, and it was on this collegiate base that the "instructors and visiting naturalists have made collections and ecological studies."[42] The significant research output of the Mountain Laboratory, which will be discussed below, emerged along with vibrant teaching. "It was found that the courses which had been planned were so well adapted to conditions at the laboratory," commented Ramaley, "and they were carried out to the satisfaction of all concerned. Students and instructors worked together on a number of special problems as well as in regular class work."[43] As the station developed, being admitted to the summer school required that students "have already pursued college courses in botany or zoology." Beginning instruction was relegated to the regular campus in Boulder, so that "chiefly ecological and systematic" fieldwork could be pursued at Tolland, including not only field and laboratory activities but also "the necessary lecture and reference work." And it was not only the instructors and visiting scholars who did serious research but sometimes even the undergraduate students themselves. "Besides students doing regular class work," the Mountain Lab's leaders noted, "a few each year take up special problems."[44] Having advanced students who undertook research work at Tolland was an aspiration of Ramaley's from the very beginning of the Lab.[45]

The Alpine Laboratory at Minnehaha initially aimed its teaching even higher. Clements focused mainly on graduate study, accepting only "advanced students in botany or related subjects, such as forestry and agronomy . . . provided they have had sufficient training to enable them to work on individual problems under adequate supervision." He foresaw the Alpine Lab as a training ground for "the methods and outlook of exact ecology," which he thought was best taught in the field. Despite the advanced education required and the independent nature of the research intended, it would still effectively function (in its early years) as a summer school, offering credit for a "full equivalent of a semester's work for the master's or the doctor's degree at the University of Minnesota, and the University of Nebraska."[46] Thus it was a different kind of pedagogy, even more advanced perhaps than what Ramaley and his colleagues at Tolland were able to muster—and certainly far higher in status than the typical kind of teachers' education that had traditionally been provided at some seaside stations—but it was nonetheless a research practice built firmly and integrally on a pedagogical base.

While the most important students for Rocky Mountain field stations were typically undergraduates from the sponsoring institutions and others, they were not the only audience for using these locations as field sites. For instance, despite its primary emphasis on teaching students from its parent institution, the University of Colorado, or from other institutions, the Mountain Lab also made a special effort to attract schoolteachers to a three-week special program scheduled right after the National Education Association meeting in Denver in 1909. This course was designed to acquaint students "with the natural history of the foothills, mountains & alpine peaks," and also included studies on plants, insects, and birds.[47] Teachers were attracted with the promise of opportunities for making collections to take back with them. The program at Tolland, Colorado, would provide for "intimate contact with nature and an acquaintence [*sic*] with mountain plants and animals," promised Ramaley.[48] The Lab charged a fee of about ten or fifteen dollars for this three-week session.[49] In some cases, then, teachers seeking continuing education in the field briefly joined the student base at Rocky Mountain field stations.

There was at least one station in the Rocky Mountains that was exclusively focused on research. But its peculiarity and late founding make it the proverbial exception that proves the rule. The Mount Evans Lab, jointly sponsored by MIT and the University of Denver and located on the summit of one of the Front Range's few peaks over fourteen thousand feet above sea level, was constructed in 1936 and opened for scientific work in 1937. It was exceptional not only in its location on an actual alpine summit and in its solely research focus, but also in its emphasis on physical rather than biological research, including studies of cosmic rays and other special high-altitude projects. Such a location was not suitable for housing students in summer courses—or, indeed, for researchers studying anything that could be studied elsewhere—due to the "rigors of the climate" that prevailed in such a place. Before the opening of the sturdier Mount Evans Lab building, it was especially dangerous on the summit. As one article about the Lab stated, "A trick on the night shift was equivalent to a polar expedition. . . . The wind velocity at night was often sufficient to level tents and scatter equipment. The fire hazard prevented safe heating of tents, and the indoor temperatures often fell to 30 degrees F. or lower. Both apparatus and workers were without protection from the frequent electric storms."[50] Even after the small Lab building was completed, it would have been an unappealing field base for anyone but those researchers whose specialties demanded it. Despite the existence of an automobile road to the summit—the highest road in the entire United States—and its proximity to Denver, which was only sixty-five miles away, it was an extremely expensive site to operate due to its forbidding location.[51] And it was really not a very good place to access a *variety* of environments, as were the stations lower down the Rocky Mountain slopes.

The Mount Evans Lab may have been an exceptionally expensive field station to operate, but many other Rocky Mountain stations were not cheap either. Bringing in supplies to remote locations high in sparsely settled mountain ranges was often difficult. The University of Wyoming's Summer Science Camp had weekly fees of fourteen dollars in the early 1940s, placing it among the most expensive field stations in the world. Such high fees were attributed to its remote location in the Medicine Bow Mountains.[52] The expense of travel during fieldwork could raise the cost even more. For students at the Summer Science Camp in 1937, for example, courses cost between $102.50 and $133.50 "depending upon the number of miles traveled during the course."[53] The long travel distances required to reach these sites also added to the expense of participants, if they were coming from institutions farther away. Still, the Wyoming Camp welcomed graduate student and faculty researchers at a per diem rate "to carry on original studies or make collections," and claimed to be able to "accommodate adequately one hundred students and staff members." Judging by the presence of ninety people, coming from nineteen institutions, during the 1936 season, the expense was no great deterrent.[54]

Whatever the independent work that was conducted by visiting researchers, the heart of a field station's work was usually its research-based curriculum. The courses of study undertaken at the Rocky Mountain stations were typically designed to train students in the latest research methods and simultaneously to undertake original studies using those methods. The New Mexico Biological Station had "original research" as its first object, followed by "the instruction of students in the methods of research," thus neatly indicating both the focus on advanced inquiry and its founders' view of the symbiosis of teaching and original research.[55] For the Alpine Laboratory, Clements outlined "four general divisions" of work for studying botany in the field, including quantitative methods, ecological methods, measurement of individual response to habitat, and quadrat studies of plant formations (see below).[56] The Mountain Laboratory stressed the study of ecology along with plant anatomy and taxonomy. Ramaley and his colleagues wanted the Tolland station to become "a center for botanical research as well as for instruction." The station's instructors undertook "ecological studies of lakes and other interesting formations," and assigned special ecology projects to their students. They described the opportunities for original research as "almost unlimited in this virgin western country," noting that "subjects worth investigating are found on every hand." Even for more conventional work in the natural history tradition, they pointed out the opportunities for original and publishable research. "Hitherto undescribed forms are continually being found," the Tolland station's leaders claimed, "and the known distributional limits of species are continually being enlarged."[57]

In addition to extending scientific knowledge, the Mountain Lab also educated students through fieldwork in the essential practices of identification, collection, preservation, and classification. The university had provided the site "for instruction in mountain zoology and botany," and especially in "the outdoor side" of those disciplines.[58] As Ramaley put it, students would not only have the opportunity to study a wide variety of natural objects around Tolland itself—plants, insects, mammals, and birds—but they would also have the chance "to become familiar with the plants of the foothills, mountains and alpine peaks."[59] In the specific outlines, then, the curriculum was also tied to the variety of environments accessible from the site. This dual focus on training students in field methods, along with providing opportunities for original research, continued in the later incarnation of the University of Colorado's summer field station as the Science Lodge in the 1920s. The Science Lodge not only offered courses in botany, ornithology, limnology, ecology, and other fields, but it also gave advanced students who wanted to do research the chance "to engage in a special problem under the supervision of the staff."[60]

Research in the Rockies

Clearly the Mountain Lab and its peers constructed an ambitious pedagogy connecting teaching goals to field experience. But how much original research did the Rocky Mountain field stations accomplish in their early years? Judged on the basis of quantity, the sites quickly did often become productive outposts for the knowledge making of their associated universities. Within its first six summer seasons, for example, the University of Montana's station at Flathead Lake had produced "360 printed pages of close type" based on work conducted at the field station.[61] Likewise, the environment of the South Boulder Creek high mountain valley around the Mountain Lab provided the setting for a relatively high proportion of field biology publications by scholars associated with the University of Colorado. The field station became in effect an extension of the campus itself into the diverse environments of the mountain zone west of Boulder, and field research conducted there came to constitute the core of the scientific papers in the *University of Colorado Studies*.

The research program of the Mountain Lab under Ramaley, as an example of fieldwork sponsored by a regional institution and leveraging place on a smaller scale, can be contrasted with another, larger-scale research program in field botany in the early twentieth century. Starting in 1900, Per Axel Rydberg of the New York Botanical Garden published his "Studies on the Rocky Mountain Flora," which transmuted in 1913 into a series of "Phytogeographical Notes on the Rocky Mountain Region." If Ramaley stressed the broad diversity of environments along the Front Range mountain slope accessible within a small radius near the Mountain Lab, Rydberg adopted a macro-scale perspective that took in the entire Rocky Mountain region from Canada south to New Mexico. Eschewing the more precise and rigorous methods of quadrats initiated by Clements and adopted by Ramaley for smaller areas, Rydberg traced the presence or absence of plant species over vast areas based on his extensive field collecting work throughout the region. In asserting that a single person could know the plants of the entire Rocky Mountain region well enough without using the rigorous, painstaking field methods of the new ecologists, Rydberg was risking their opprobrium—and, indeed, he often prefaced his "Phytogeographical Notes" with anticipatory defensive comments.[62] These were two different strategies that implied two contrasting notions of how to extend place beyond the local.

In contrast to Rydberg's singlehanded pursuit of a synoptic view of the distribution of plants across the Rocky Mountains, the publishing researchers at Ramaley's Mountain Lab were more diverse. The most productive researcher at Tolland, in terms of publications, was Ramaley himself, who began by publishing a write-up of notes from his July 1909 lectures at Mountain Lab.[63] Shortly thereafter, student work from the 1909 season led to publication as part of what was called the Colorado Biological Survey (CBS). The authors include two men and one woman, all receiving B.A. degrees soon afterward.[64]

C. H. Edmondson, a professor at Washburn College in Topeka, Kansas, also published an article in the CBS series based on 1911 summer fieldwork in counties near the Mountain Lab.[65] In addition, students at the Mountain Lab assisted a larger, statewide natural history cataloging project by University of Colorado biology instructor Wilfred Robbins, itself a part of the CBS.[66] Moreover, biological materials collected by Edmondson, Ramaley, and Robbins around Tolland were later described by students back in Boulder in the CBS series.[67] Finally, specimens obtained over several years, such as the vascular plants collected "chiefly in connection with ecological studies" by Ramaley (plants in "dry grassland, meadows and forests") and Robbins ("marsh and streamside vegetation"), were later described at great length in print with the taxonomic help of professor Aven Nelson of the nearby University of Wyoming.[68]

In addition to articles in the university's own journal, publication based on Mountain Lab research appeared in national journals such as the *Botanical Gazette, Plant World*, and the *Bulletin of the Torrey Botanical Club*.[69] While the science done at Tolland lacked the international stature of Clements's work at the Alpine Lab—which was widely regarded as foundational in advancing work on ecological problems—it nevertheless did enable researchers associated with the Mountain Lab to produce knowledge that would circulate in leading national scientific journals.[70] These studies from a wide assortment of researchers who worked at the Mountain Lab spanned a variety of field sciences, from well-known specialties such as botany and entomology, to less-famous ones such as conchology (the study of mollusk shells) and bryology (the study of mosses). Such work was often conducted in the traditional natural history mode, as when Mary Esther Elder wrote up a plant collection made by Ramaley in the summer of 1911 along the Rollins Pass Road near Tolland, or when Texas A&M's E. L. Reed studied the butterflies around the Mountain Lab as part of his graduate school work, or when Ramaley's University of Colorado colleague T. D. A. Cockerell found a higher-than-usual specimen of the sagebrush mealybug and several species of snails, including one new to the Colorado list, when collecting at Tolland with his wife in August 1911.[71]

Likewise, A. J. Grout, a bryologist who taught at Curtis High School in New York City, spent the summer of 1914 at the Mountain Lab, realizing a longtime "desire to collect and study the mosses of the Rocky Mountains." He was assisted in his moss collecting by "Miss H. A. Leonard, a student at the summer school," thus once again reinforcing the symbiosis of the teaching and research practices at Tolland. Grout was especially grateful that Leonard "collected in localities I could not reach and at other seasons of the year." Based on their combined collections, Grout determined the mosses on the eastern slope of the Rocky Mountains to be "more Eastern than Pacific," which seems to be a somewhat surprising conclusion.[72] In a small and very specialized discipline such as bryology, the Mountain Lab was critical to gaining access to a new and relatively unstudied environmental region. While most of these visiting researchers only came for one field season, they were a crucial part of the Mountain Lab's activities.

But it was Ramaley himself who conducted the most studies at the Mountain Lab and the ones involving the most systematic research. In a move that suggests his intent to make his field research more rigorous and compelling according to cosmopolitan scientific standards, Ramaley adopted the use of quadrats, which had been introduced by Clements (and fellow Nebraskan Roscoe Pound) a few years earlier to quantitatively measure vegetation in species composition and changes over time within a fixed square space. Beginning with 19 quadrats in the vicinity of Tolland in the summer of 1911, Ramaley continued to deploy the quadrat as a tool for rigorous field study, delineating up to 158 meter-sized quadrats in the summers of 1914 and 1915. This quadrat-based research led to a string of publications in a leading national journal, the *Botanical Gazette*.[73] From available evidence, it appears that most, if not all, of Ramaley's quadrats were in the mountain valley around Tolland, rather than spread out from the foothills to the alpine peaks.

Ramaley's work in the Mountain Lab's valley environment also led to a rather lengthy paper in *Plant World*, a journal that was transformed during the 1910s from a more traditional natural history focus on lists and descriptive botany to a rigorous journal of highly technical, quantitative ecology under the leadership of scientists affiliated with the Desert Botanical Laboratory in Arizona. Ramaley's research for that paper began in 1912, with the "most intensive" work carried out "during the three summers from 1913 to 1915 inclusive."[74] A few years later, Ramaley published papers in the *American Journal of Botany*—a new national journal begun in 1914 with the support of the Botanical Society of America—based on studies at the Mountain Lab on ecological relationships as well as an extended study of subalpine lakes across the region, making use of vegetation maps and quadrats.[75] During the 1910s, Ramaley also invented a "Soil Moisture Index," which he described in the *Botanical Gazette* as rooted in his teaching practice in the field around the Mountain Lab. "A rule of the [Tolland] laboratory," explained Ramaley, "required that every plant studied should be recorded with an index number," in order for students to learn about associations between plants and soil moisture.[76] Thus Ramaley's publication record demonstrates not only how he used the summer school base to build his own research program but also the continuing integration of advanced pedagogy and rigorous field science practices at the Mountain Lab.

Ultimately, however, it is hard to miss the striking concentration of the Mountain Lab's research program, especially the most systematic and rigorous work led by Ramaley himself, in the high mountain valley around Tolland known as Boulder Park. So what became of the highly touted wide range of environments from plains and foothills to alpine peaks? Was that merely part of the Mountain Lab's teaching mission (a presumably difficult distinction given the careful integration of the teaching and research that prevailed at the station)? Were there spatial limits to the bridging of the epistemic rift between residential experience and cosmopolitan knowledge making? In practice, it appears that the high intensity of investment required to produce rigorous research using instruments, numerical indices, and quadrats made it difficult, in practice, to extend research to the full range of environments around the station. Similarly, but more frankly, the Rocky Mountain Biological Laboratory at Gothic "center[ed] its activities around problems of the alpine and sub-alpine regions."[77]

At the Mountain Lab, a rare exception to the focus on Boulder Park was an article that Ramaley published in *Bulletin of the Torrey Botanical Club* in 1919, in which he compared xerophytic (dry) grasslands from the mountain front mesa area at 5,340 feet, through the valley around Tolland at 8,889 feet, to the subalpine heights of Bryan Mountain at 11,000 feet. Yet this study, based on ten years of observations, revealingly eschewed the more systematic methods of quadrats and quantification in favor of a four-category classification of each species at each altitude as "present in dry grassland," "found in moister places rather than in dry grassland," "occurs . . . but only in special situations (where warm or protected)," or completely absent.[78] While not on the same large scale as Rydberg's studies of the entire Rocky Mountain region, it shared a partial resemblance to them. Achieving greater spatial scale, it seems, also entailed abandoning the more intensive and stringent field research methodologies employed around Tolland itself.

Such trade-offs between rigor and scale suggest the ultimate limitations of the early twentieth-century strategy to make local field stations produce global knowledge—to make the slopes of the Rocky Mountains stand for the whole of the North American continent. Perhaps in recognition of the ultimate limitations of summer field stations, in the Rockies or elsewhere, to produce cutting-edge research, the Rockefeller Foundation in the 1920s summarily turned down a funding request from the Rocky Mountain Biological Lab at Gothic. "While it is true that we have given some assistance to the biological stations at Woods Hole and at Pacific Grove," acknowledged the Foundation officer, referring to the Massachusetts station mentioned above and another pioneering seaside station located near Monterey, California, "this has been because of the national, or even international, significance of these laboratories. I am afraid our Board would not be willing to consider aid to a comparatively small laboratory maintained by a single college unless we were prepared to undertake support of such summer courses on a very wide basis."[79] The kind of science undertaken at the Rocky Mountain field stations in the early twentieth century, with few exceptions, ultimately failed to achieve a high enough status to attract competitive national funding. It was high science, perhaps, but not high enough.

Conclusion

This essay has considered the early work of Rocky Mountain field stations, concentrating especially on the University of Colorado's field station in the small village of Tolland. Located in a high, glaciated valley below tree line, it operated from 1909 to 1919 in the local context of its recreational resort village high in the Front Range of the Rocky Mountains, the regional context of its distinctiveness as an advanced research and teaching outpost of a state university in the interior West, and the quasi-global context of its aspirations to encompass a broad diversity of environmental zones spanning practically the entire temperate world. Most stations in the interior West can be closely tied to the region's practical, applied concerns, but field stations in the Rocky Mountains show how nature could also be harnessed for the "high science" of rigorous methods and advanced research training.

Yet, in the end, it is clear that the Rocky Mountain stations were only partially successful in the geographical ambitions of their high-science aims. Research output, and even teaching, ultimately focused predominantly on the mountain valleys in which the stations themselves were located, rather than reflecting the broad diversity of places accessible by moving vertically between mountaintop and plain. In other words, there were limitations to how easily a field station could function to bridge the epistemic rift between residential experience and cosmopolitan knowledge making: the two could be combined, but only so far. Beyond the immediate environs of the field station—in the case of the Mountain Lab, the glaciated high mountain valley around Tolland known as Boulder Park—the rich synergy of pedagogy and original inquiry rapidly attenuated. Nevertheless, by showing how the Mountain Lab integrated teaching and research, I hope to have illuminated how field stations in the Rocky Mountains attempted to expand the geographical scale of knowledge, both in principle and practice.

The key to the attraction of the Rocky Mountain field station was its distinctive combination of recreational amenities with the prospect for serious research and advanced teaching across a wide range of environmental zones. While this ideal may not have been fully realized, it permeated both the rhetoric and practice of stations such as the Mountain Laboratory of the University of Colorado. In promoting the University of Montana's field station on Flathead Lake in 1904, Morton Elrod captured succinctly the combination of recreation with advanced research: "One can recreate to the fullest and yet have a definite object in view."[80] By harnessing the place-based features of the Rocky Mountains to combine ongoing daily experiences of the natural world with cosmopolitan ambitions to produce knowledge, field scientists working there attempted to not only pursue a "definite object" of high-level research but also teach advanced research methods to students. Rather than grounding original research in the economic development context of agriculture, mining, ranching, or other crucial activities that prevailed in the region at this time, the early Rocky Mountain field stations largely aimed for a practical symbiosis with pedagogy.

Notes

1. For example, see the recent work of University of Maryland scientist David Inouye at the Rocky Mountain Biological Laboratory in Gothic, Colorado, as described on the Laboratory's website, "Plants Point to Global Warming," Science in Action: Featured Projects, Rocky Mountain Biological Laboratory, http://rmbl.org/rockymountainbiolab/science.html.
2. My concept of "epistemic rift" is deliberately modeled on the concept of "metabolic rift," which has been revived as a conceptual tool for environmental history in Jason W. Moore, "Environmental Crises and the Metabolic Rift in World-Historical Perspective," *Organization and Environment* 13 (2000): 123–157. The associated terminology of "residential" and "cosmopolitan" is derived from Robert Kohler's work, as noted in the introduction to this volume.
3. Francis Ramaley, "The University of Colorado Mountain Laboratory," *University of Colorado Studies* 7 (1909): 91–95, on p. 91.
4. According to Robert E. Kohler, who coined the term, "labscapes" were places set up by "altering the physical design and siting of labs, making their interiors more natural and lowering their thresholds with the world of nature." Kohler, "Labscapes: Naturalizing the Lab," *History of Science* 40 (2002): 473–501, on p. 474.
5. On how knowledge is linked to local place in a quintessentially practical field science, agricultural extension, see Christopher R. Henke, "Making a Place for Science: The Field Trial," *Social Studies of Science* 30 (2000): 483–511.
6. The same pattern had developed earlier at U.S. marine stations, where the American model of combining teaching with advanced research prevailed. See Keith R. Benson, "Laboratories on the New England Shore: The 'Somewhat Different Direction' of American Marine Biology," *New England Quarterly* 61 (1988): 55–78. On the central role of teaching in the development of laboratories, see Graeme Gooday, "Precision Measurement and the Genesis of Physics Laboratories in Victorian Britain," *British Journal for the History of Science* 23 (1990): 25–51.
7. Robert E. Kohler, "The Ph.D. Machine: Building on the Collegiate Base," *Isis* 81 (1990): 638–662.

8. H. A. Rowland, "A Plea for Pure Science," *Science* 2 (1883): 242–250, on p. 246. There is an extensive secondary literature on the moral values attached to higher education; for example, see Larry Owens, "Pure and Sound Government: Laboratories, Gymnasia, and Playing Fields in Nineteenth-Century America," *Isis* 76 (1985): 182–194.
9. On Woods Hole, see Philip J. Pauly, "Summer Resort and Scientific Discipline: Woods Hole and the Structure of American Biology, 1882–1925," in *The American Development of Biology*, ed. Ronald Rainger, Keith R. Benson, and Jane Maienschein (Philadelphia: University of Pennsylvania Press, 1988), 121–150; and Jane Maienschein, "History of American Marine Laboratories: Why Do Research at the Seashore?" *American Zoologist* 28 (1988): 15–25.
10. Kohler, "Labscapes," 476, 480–481. In this regard, the mountain field stations discussed in this paper, with their tilt toward the life sciences, present a chronology quite distinct from that of the "mountain science" discussed in a recent special issue of *Science in Context*, which considers the earth and physical sciences and charts a late eighteenth-century rise followed by an early twentieth-century decline, thus bracketing the "high age of Alpine mountaineering." See Charlotte Bigg, David Aubin, and Philipp Felsch, "Introduction: The Laboratory of Nature—Science in the Mountains," *Science in Context* 22 (2009): 311–321, on p. 313.
11. Sharon E. Kingsland, "An Elusive Science: Ecological Enterprise in the Southwestern United States," in *Science and Nature: Essays in the History of the Environmental Sciences*, ed. Michael Shortland (Oxford: British Society for the History of Science, 1993), 151–179; and Kingsland, *The Evolution of American Ecology, 1890–2000* (Baltimore: Johns Hopkins University Press, 2005). For a global inventory of biological field stations, see Homer A. Jack, "Biological Field Stations of the World," *Chronica Botanica* 9 (1945): 1–73.
12. Kohler, *Landscapes and Labscapes: Exploring the Lab-Field Border in Biology* (Chicago: University of Chicago Press, 2002), 51–55.
13. On early Rocky Mountain exploration and surveys, see William Goetzmann, *Exploration and Empire: The Explorer and the Scientist in the Winning of the American West* (New York: Knopf, 1966); and Thomas G. Manning, *Government in Science: The U.S. Geological Survey, 1867–1894* (Lexington: University of Kentucky Press, 1967).
14. P. A. Welker, "Electric Storm on Mount Elbert, Colorado," *Science* 2 (1895): 304–306.
15. Rebecca Robbins Raines, *Getting the Message Through: A Branch History of the U.S. Army Signal Corps* (Washington, D.C.: Center of Military History, 1996), 49; Phyllis Smith, *Weather Pioneers: The Signal Corps Station at Pikes Peak* (Athens, Ohio: Swallow Press, 1993); and Donald R. Whitnah, *A History of the United States Weather Bureau* (Urbana: University of Illinois Press, 1961). On the decision not to locate an astronomical observatory on Pikes Peak, see Catherine Nisbett Becker, "Professionals on the Peak," *Science in Context* 22 (2009): 487–507.
16. One precursor was the short course offered at the Colorado Summer School of Science, Philosophy, and Language, which included some outdoor botanical instruction, held in Colorado Springs in the summer of 1895, as noted in Charles E. Bessey, "Summer-School Botany in the Mountains," *American Naturalist* 29 (1895): 845–847.
17. On the role of Clements, his colleagues, and his students in the origins of American ecology, see Ronald C. Tobey, *Saving the Prairies: The Life Cycle of the Founding*

School of American Plant Ecology, 1895–1955 (Berkeley and Los Angeles: University of California Press, 1981). See also Joel B. Hagen, "Clementsian Ecologists: The Internal Dynamics of a Research School," *Osiris* 8 (1993): 178–195. For details of the rise and fall of the Alpine Laboratory as an object of funding by the CIW, see Patricia Craig, *Centennial History of the Carnegie Institution of Washington*, vol. 4, *The Department of Plant Biology* (Cambridge: Cambridge University Press, 2005), 34–35, 98.

18. Frederick E. Clements, "The Alpine Laboratory," *Science* 37 (1913): 327.
19. "Mountain and Alpine Laboratories, Colorado," *Journal of Ecology* 1 (1913): 157.
20. "Location" (n.d.), Morton J. Elrod Papers, K. Ross Toole Archives, University of Montana, Missoula, Box 28, Folder 17.
21. "The Alpine Laboratory of the Botanical Seminar of the University of Nebraska," *Science* 20 (1904): 185. Evocatively, the author (obviously Clements) continues: "The dream of the physiologist (ecologist) to have his laboratory out-of-doors may be realized here, and it is merely a matter of time until methods will be found by which research will deal primarily with the experiments of nature, and the walled laboratory will be relegated to a purely secondary place."
22. Francis Ramaley and W. W. Robbins, "A Summer Laboratory for Mountain Botany," *Plant World* 12 (1909): 105–110, on p. 107. For Ramaley's synopsis of the five vegetation zones of the Rockies, published just before he founded the Mountain Laboratory, see Ramaley, "Plant Zones in the Rocky Mountains of Colorado," *Science* 26 (1907): 642–643.
23. Hugo G. Rodeck, "Science Lodge," *Biologist* 18 (1937): 101. For a similar statement about diverse life zones accessible at one of the earliest Rocky Mountain field stations, see T. D. A. Cockerell, "The New Mexico Biological Station," *Science* 13 (1901): 954.
24. Kohler, *Landscapes and Labscapes*, 164–165. Clements eventually used mountain slopes not only to stand for large geographical spaces but also to probe backward and forward in time through successional changes in biomes resulting from climate change. See Clements, "The Relict Method in Dynamic Ecology," *Journal of Ecology* 22 (1934): 39–68, esp. p. 49, and analyzed in Kohler, *Landscapes and Labscapes*, 225–226.
25. John C. Johnson, "Rocky Mountain Biological Laboratory," *Biologist* 18 (1937): 105; and "Opportunities for Biological Research and Field Study," undated typescript, John C. Johnson Collection, Leslie J. Savage Library, Western State College of Colorado, Gunnison, Box 2.
26. "Science Summer Camp," *Biologist* 18 (1937): 183–184.
27. Ramaley, "University of Colorado Mountain Laboratory," 94–95.
28. "Opportunities for Biological Research," n.p.
29. Robert E. Kohler, *All Creatures: Naturalists, Collectors, and Biodiversity, 1850–1950* (Princeton: Princeton University Press, 2006), esp. chap. 2. See also Kohler, "Labscapes," 484–485.
30. Francis Ramaley and G. S. Dodds, "University of Colorado Mountain Laboratory," *University of Colorado Studies* 12 (1917): 53–64, on pp. 53–54.
31. Ramaley and Robbins, "Summer Laboratory," 105.
32. Ibid., 106–107.
33. *Vacation and Study in the Heart of the Rockies*, 1931 summer bulletin of Western State College of Colorado, available in Leslie J. Savage Library, Western State College of Colorado, Gunnison.

34. "School of Mountain Field Biology of the University of Colorado" (n.d.), Francis Ramaley Papers, University of Colorado, Boulder, Archives (hereafter cited as FRP), Box 1, Folder 26.
35. Ramaley and Robbins, "Summer Laboratory," 109.
36. "Science Summer Camp," 183.
37. Rodeck, "Science Lodge," 102.
38. Martin J. Elrod, "The University of Montana Biological Station," *Journal of Applied Microscopy and Laboratory Methods* 4 (1901): 1269–1270. For a broader discussion of the relationship between railroads and scientific fieldwork, see Jeremy Vetter, "Science Along the Railroad: Expanding Field Work in the US Central West," *Annals of Science* 61 (2004): 187–211.
39. Ramaley and Robbins, "Summer Laboratory," 108.
40. On the importance of teaching in the genesis of American marine biological stations, see Keith R. Benson, "Why American Marine Stations? The Teaching Argument," *American Zoologist* 28 (1988): 7–14.
41. Jack, "Biological Field Stations," 8–9. See also Kohler, "Labscapes," 484.
42. Ramaley and Dodds, "University of Colorado Mountain Laboratory," 53.
43. Ramaley, "University of Colorado Mountain Laboratory," 91.
44. Ramaley and Dodds, "University of Colorado Mountain Laboratory," 53.
45. Ramaley, "University of Colorado Mountain Laboratory," 94.
46. Clements, "Alpine Laboratory," 328.
47. "School of Mountain Field Biology of the University of Colorado" (n.d.), FRP, Box 1, Folder 26.
48. Ramaley, "University of Colorado Mountain Laboratory," 94.
49. Ramaley to Mr. Fred E. Hagen, 3 July 1909, FRP, Box 1, Folder 9.
50. J. C. Stearns, "The Mount Evans Laboratory," *Scientific Monthly* 46 (1938): 242.
51. According to Homer Jack's survey, in fact, it had the most expensive user fee of any biological station in the world in the mid-1940s, at one hundred dollars per month, as noted in Jack, "Biological Field Stations," 33.
52. Ibid., 21.
53. "Science Summer Camp," 188.
54. Ibid., 186.
55. J.A. Schufle, *Fragments from the Stream of Time: A History of the Sciences at New Mexico Highlands University* (Las Vegas: New Mexico Highlands University, 1968), 4.
56. Clements, "Alpine Laboratory," 327.
57. Ramaley and Robbins, "Summer Laboratory," 108, 110.
58. Ramaley and Dodds, "University of Colorado Mountain Laboratory," 53; and Ramaley, "University of Colorado Mountain Laboratory," 94.
59. Ramaley to Mr. Fred E. Hagen, 3 July 1909, FRP, Box 1, Folder 9. See also Ramaley, "University of Colorado Mountain Laboratory," 91.
60. Rodeck, "Science Lodge," 103.
61. Elrod, "University of Montana Biological Station," 206. It should be noted that although Elrod touted the favorable opportunities for "carrying on work in plant ecology . . . due to the fact that within a radius of ten miles there are several different climates, giving rise to different plant formations" (207), much of the research output was done in the traditional natural history mode, including the work of ornithologist and oologist P. M. Silloway, who made his living as a teacher at Fergus County High

School in Lewistown, Montana. Silloway's papers include "Flathead Lake Findings," *Condor* 3 (1901): 4–7; "The Holboell Grebe in Montana," *Condor* 4 (1902): 128–131; "Afield at Flathead," *Condor* 6 (1904): 12–14; "Notes from Flathead," *Condor* 7 (1905): 19–22; "The American Crossbill in Montana," *Condor* 7 (1905): 174–176; and "Among the Flathead Birds," *Condor* 8 (1906): 109–110.

62. Contrast Rydberg's attempt to increase the geographical scope of his research with the more involved and rigorous research program of Frederic Clements to scale up his ecological methods to "universal application" during the 1913 and 1914 field seasons by covering an even larger region, "from the Great Plains to the Pacific coast, and from the Canadian Rockies to the Mexican boundary," as noted in Clements, *Plant Succession: An Analysis of the Development of Vegetation* (Washington, D.C.: Carnegie Institution of Washington, 1916), iii. Clements's comments suggest the high level of rigorous method involved. "The great climax formations of this region were traversed repeatedly," he noted, "and their development and relations subjected to critical analysis and comparison."
63. Francis Ramaley, "Remarks on Some Northern Colorado Plant Communities with Special Reference to Boulder Park (Tolland, Colorado)," *University of Colorado Studies* 7 (1910): 223–236.
64. Dean T. Prosser, "Habits of 'Amblystoma Tigrinum' at Tolland, Colorado," *University of Colorado Studies* 8 (1911): 257–263; Katharine Bruderlin, "A Study of the Lodgepole-Pine Forests of Boulder Park (Tolland, Colorado)," *University of Colorado Studies* 8 (1911): 265–275; and Francis Ramaley and Louis A. Mitchell, "Ecological Cross-Section of Boulder Park (Tolland, Colorado)," *University of Colorado Studies* 8 (1911): 277–287.
65. C. H. Edmondson, "Protozoa of High Mountain Lakes in Colorado," *University of Colorado Studies* 9 (1912): 65–74.
66. Wilfred W. Robbins, "Preliminary List of the Algae of Colorado," *University of Colorado Studies* 9 (1912): 105–118.
67. G. H. Wailes, "Some Desmids from Alpine Stations in Colorado," *University of Colorado Studies* 9 (1912): 119–120; and Francis Ramaley and Mary Esther Elder, "The Grass-Flora of Tolland," *University of Colorado Studies* 9 (1912): 121–141.
68. Francis Ramaley, "Vascular Plants of the Tolland Region in Colorado," *University of Colorado Studies* 12 (1917): 27–51, on pp. 27–28.
69. For a complete list, see Ramaley and Dodds, "University of Colorado Mountain Laboratory," 63–64.
70. On Clements's own self-assessment of the cosmopolitan significance of work at the Alpine Lab, see Clements, "Alpine Laboratory," 327. See also Clements, *Research Methods in Ecology* (Lincoln, Neb.: University Publishing Company, 1905).
71. Mary Esther Elder, "Roadside Plants of a High Mountain Park in Colorado," *Torreya* 12 (1912): 175–180; E. L. Reed, "Butterflies of a Mountain Park in Colorado (*Lep.*)," *Entomological News* 27 (1916): 267–268; T. D. A. Cockerell, "Eriococcus borealis in Colorado," *Journal of Economic Entomology* 5 (1912): 295; and Cockerell, "Land Mollusca at Tolland, Colorado," *Nautilus* 25 (1911): 58–59.
72. A. J. Grout, "Mosses of Colorado from Tolland and Vicinity," *Bryologist* 19 (1916): 1–8, on pp. 1, 2.
73. Francis Ramaley, "The Amount of Bare Ground in Some Mountain Grasslands," *Botanical Gazette* 57 (1914): 526–528; Ramaley, "The Relative Importance of Different

Species in a Mountain Grassland," *Botanical Gazette* 60 (1915): 154–157; and Ramaley, "Quadrat Studies in a Mountain Grassland," *Botanical Gazette* 62 (1916): 70–74.

74. Francis Ramaley, "Dry Grassland of a High Mountain Park in Northern Colorado," *Plant World* 19 (1916): 249–270, on p. 249.
75. Francis Ramaley, "The Role of Sedges in Some Colorado Plant Communities," *American Journal of Botany* 6 (1919): 120–130; and Ramaley, "Subalpine Lake-Shore Vegetation in North-Central Colorado," *American Journal of Botany* 7 (1920): 57–74.
76. Francis Ramaley, "Soil Moisture Index," *Botanical Gazette* 63 (1917): 151–152.
77. Johnson, "Rocky Mountain Biological Laboratory," 106.
78. Francis Ramaley, "Xerophytic Grasslands at Different Altitudes in Colorado," *Bulletin of the Torrey Botanical Club* 46 (1919): 42.
79. George E. Vincent to Dr. Quigley, 22 September 1926, John C. Johnson Collection, Leslie J. Savage Library, Western State College of Colorado, Gunnison, Box 2.
80. Morton J. Elrod, "The University of Montana Biological Station and Its Work," *Science* 20 (1904): 205–212, on p. 212.

Six

Mark V. Barrow Jr.

On the Trail of the Ivory-Bill

Field Science, Local Knowledge, and the Struggle to Save Endangered Species

In early January 1937, James T. Tanner loaded up his car and headed south. Over the next three years the twenty-two-year-old Cornell graduate student would spend nearly twenty-one months in the field, logging more than forty thousand miles on his 1931 Model A Ford and covering untold additional terrain by train, foot, boat, and horse.[1] The goal of this long, arduous journey was to learn more about the elusive ivory-billed woodpecker, known to science as *Campephilus principalis*, but known variously to locals across its range as the "ivory-bill," the "kent," the "king of the woodpeckers," and the "Lord God bird." With its striking coloration, curious call, and wingspan of approximately thirty inches, it was not only the largest woodpecker in the United States but also one of the most impressive of its kind anywhere in the world. This awe-inspiring bird had once graced the moist bottomlands and swampy forests across the Southeast. But as its habitat increasingly fell victim to agricultural development and logging in the years surrounding the American Civil War, its range had been dramatically reduced and its numbers dangerously thinned.[2] Although occasional claims of sightings continued at a handful of locations, by the time Tanner departed from Ithaca, it was clearly teetering on the brink of extinction. He hoped to discover exactly how many ivory-bills remained, to identify the specific sites they still inhabited, and to document their life history, behavior, and ecology. Ultimately Tanner and his sponsor, the National Association of Audubon Societies, hoped his research would lead to policies that would rescue the beleaguered species.

As the most thorough scientific investigation of an endangered species conducted up to that point, Tanner's study came at a crucial juncture in the development of American wildlife management. Within a few short years, amid the Great Depression, the naturalist Herbert Stoddard published the first American monograph on the biology and management of a game species (1931); the forester-turned-wildlife-specialist Aldo Leopold not only authored the first textbook but also secured the first university chair specifically devoted to the field (1933); the U.S. Bureau of Biological Survey established a Cooperative Wildlife Research Unit Program in conjunction with eight American universities (1935); and wildlife managers and conservationists convened the first North American Wildlife Conference (1936), founded their own professional organization, The Wildlife Society (1937), and began publishing the *Journal of Wildlife Management* (1937). Accompanying this intense period of institutionalization and professionalization, a handful of wildlife managers began calling for an increased reliance on science to learn more about the factors governing the abundance and distribution of native flora and fauna. It was time, these reformers argued, to abandon the crude "rule-of-thumb" calculations that had long characterized their field. It was also time to broaden the preoccupation with game species, which had garnered the lion's share of attention and funding.[3]

Perhaps not coincidentally, calls for reform in wildlife management came not long after the field of natural history experienced a series of profound shifts of its own. Naturalists who had once learned their craft through apprenticeships and self-study were quickly giving way to university-trained biologists. As training in natural history became more formalized during the first decades of the twentieth century, the traditional focus on collecting, naming, and describing new species that had long gripped the enterprise broadened to an interest in the ecology and behavior of living organisms in the field. By 1910 the Ph.D. degree had become a standard entry credential into most scientific disciplines. Within the next decade a series of graduate programs at places like Cornell, Michigan, and Berkeley were churning out university-trained naturalists with both the scientific tools and the inclination to study native species in their natural environment.[4]

Accompanying these changes in the organization and practice of science was an escalating concern about human-induced wildlife extinction. Despite many well-publicized conservation successes, numerous native species continued to fall victim to habitat destruction, over-hunting, competition with exotics, introduced diseases, and other human-driven transformations of the landscape. By the time Tanner began his study in 1937, the great auk, the Labrador duck, and the passenger pigeon had all been clearly lost.[5] Only four years previously, conservationists had read with horror about the disappearance of Booming Ben, the final heath hen. This once ubiquitous subspecies had perished despite the fact that a Harvard-trained biologist, Alfred O. Gross, had carefully monitored the species during its final years on Martha's Vineyard, Massachusetts, and state officials spent more than seventy thousand dollars trying to rescue the beleaguered bird.[6] Numerous other species—the California condor, the whooping crane, the Carolina parakeet, the trumpeter swan, and the ivory-bill among them—seemed to be facing a similar fate.[7] Whether couched in utilitarian, philosophical, aesthetic, or cultural terms, by the first half of the twentieth century, the specter of extinction haunted a growing number of American naturalists, conservationists, and nature lovers.

Although initiated in response to escalating concerns about the fate of vanishing wildlife and recent reforms in the field of wildlife management, Tanner's study represented a curious combination of old and new approaches to natural history. To identify potential sites where remaining ivory-bill populations might persist, Tanner relied on the basic techniques naturalists had long used to compile inventories of the world's flora and fauna: he scanned the scientific literature for references to the species, examined museum specimens with collection data, and ventured into the field himself. And just as naturalists had always depended on an extended, heterogeneous community of collaborators to provide the specimens necessary for their taxonomic research, Tanner also relied on the eyes and ears of a diverse group of hunters, trappers, wardens, loggers, and other local residents to discover and evaluate promising ivory-bill habitat. Moreover, during the extensive phase of his field research, he counted on local guides to help him navigate through the remote, often difficult-to-traverse wetlands the species was known to once inhabit. During the more intensive phase of his research—when he spent three seasons observing the ivory-bill and its habitat at a single site in Louisiana—Tanner increasingly turned to the methods and practices of modern field science. But even at his main study site he continued to lean heavily on the knowledge, skills, and instincts of a local woodsman who helped him locate and monitor ivory-bills. In short, in addition to formal university training, what might be appropriately termed "indigenous" or "local" knowledge possessed by long-term residents with an intimate, prolonged experience of particular landscapes played a key role in the success of Tanner's ivory-bill study.[8]

Whatever the source of the knowledge Tanner produced, modern science did offer the hope of providing authoritative information about the status of endangered species like the ivory-bill. In fact, Tanner succeeded in locating a remnant population of the rare species, identifying the critical habitat it needed, and offering a series of policy recommendations that might have fostered its long-term survival. But as he and his sponsor, the National Association of Audubon Societies, soon discovered, while science might be necessary to save the ivory-bill, it was hardly sufficient. Also needed was the political will to implement Tanner's modest recommendations. And in the early

1940s, Americans seemed much more focused on preserving private property rights, promoting economic development, and winning the war than in saving a rare, swamp-dwelling bird that survived by eating grubs.

Early Encounters

The British naturalist Mark Catesby first described the ivory-billed woodpecker in the early eighteenth century, one of dozens of North American birds he introduced to the world. In his brief account, Catesby noted that the woodpecker's white bill was "much valued by the *Canada Indians*, who made Coronets of them for their princes and great warriors, by fixing them around a wreath, with their points outwards."[9] Nearly a century later the American artist and naturalist Alexander Wilson captured a live specimen near Wilmington, North Carolina, to serve as a model for one of his ornithological drawings. When he left the mildly wounded bird in his hotel room, it wreaked havoc with its bill in a vain attempt to escape. Wilson noted that his captive "displayed such a noble and unconquerable spirit" that he was sorely tempted to "restore him to his native woods." He failed to do so, however, and Wilson's woodpecker perished just three days later.[10]

By the first several decades of the nineteenth century, Wilson, John James Audubon, and other American naturalists had recorded the basic facts regarding the ivory-bill's appearance and habitat.[11] Its glossy black plumage was broken up by a white stripe that started on each cheek and continued down the side of its neck. It also possessed a broad band of white on the trailing edge of its wings, an area that appeared as a conspicuous triangular-shaped patch on its back when the bird was perched. Its bill was ivory-white, approximately two and a half inches long, and about an inch wide at its base. The male boasted a prominent scarlet crest, while the crest of the female was entirely black. Audubon compared the bird's call to the high false note of a clarinet, while others described it as a nasal "kent" or a child's tin trumpet. Although it was (and still is) often confused with the much more common pileated woodpecker, the ivory-bill was slightly larger, its bill was whiter, and its white wing patches were visible on its back when at rest and on the trailing edge of its wings when in flight. Early naturalists also recognized the ivory-bill's preference for deep bottomland forests in the southeastern United States.

By the end of the nineteenth century, the ivory-bill had declined to the point that it was considered one of North America's rarest birds. As such, its skins and eggs proved in great demand among the many individuals amassing private and institutional natural history collections during that period.[12] Even naturalists who were otherwise sympathetic to avian conservation rarely passed up the opportunity to acquire examples of the increasingly prized bird. For example, the ornithologists Frank Chapman and William Brewster were both founders of Audubon bird-protection movement in the late 1880s, and for several decades both remained prominent leaders in a highly successful campaign to cultivate public sympathy and legislative protection for nongame birds.[13] Despite their strong commitment to bird conservation, however, when the two naturalists spotted an ivory-bill during an 1890 expedition down Florida's Suwannee River, they expressed no qualms about collecting the rare bird.[14] The rage for collecting proved so compelling that by 1920 more than four hundred ivory-bill skins were gathering dust in ornithological collections across the world, yet naturalists had failed to record anything beyond the most rudimentary facts concerning the species' life history and behavior.[15] By that time many feared the bird might have been gone entirely.

Rediscovering the Ivory-Bill

Predictions of the ivory-bill's demise, however, proved premature. Rumors of sightings continued to circulate in ornithological circles and appear in print. In 1932 Audubon president T. Gilbert Pearson rushed to Madison Parish, Louisiana, following reports that a local attorney and state legislator had recently acquired an ivory-bill specimen.[16] To his astonishment, Pearson spotted the species on a massive tract of swampy hardwood forest owned by the Singer Manufacturing Company. In his excitement about the find, however, Pearson had carelessly mentioned to a local reporter that ivory-bill specimens might fetch as much as one thousand dollars each from eager collectors. Although the species was officially protected by the Louisiana Department of Conservation, ivory-bill defenders feared that the valuable bird would prove a tempting target to residents in this Depression-torn region of rural Louisiana.

In 1934 Arthur A. Allen, a professor of ornithology at Cornell, heard firsthand about the recently sighted Madison Parish ivory-bills while attending the annual meeting of the American Ornithologists' Union.[17] At the time, he was considering how to spend his sabbatical leave the following spring. During the previous decade, Allen had been building Cornell's ornithology and wildlife conservation program into one of the largest, most successful of its kind in the nation.[18] Cornell was also one of the moving forces behind the American nature study movement, so not surprisingly, besides his extensive teaching and research duties, Allen remained deeply committed to the popularization of ornithology.[19] During his long career he not only wrote dozens of articles for *Bird-Lore, National Geographic*, and other popular venues, but also delivered countless public lectures on birds, many of them illustrated with his own photographs and motion pictures. In the late 1920s, he helped secure some of the first successful sound recordings of birds in the wild. Joining him in these pioneering efforts were two of his students, Peter Paul Kellogg and Albert R. Brand, both of whom would devote much of their subsequent careers to recording birdsong for scientific study and popular consumption.[20]

With financial backing from Brand, who had relinquished a seat on the New York Stock Exchange to study ornithology, Allen decided to mount an expedition to record the voices of North American birds, particularly species that seemed to be vanishing.[21] A special object of the expedition was "the rediscovery of the ivory-billed woodpecker, perhaps the rarest of North American birds and at one time thought to be extinct."[22] Allen and his colleagues also hoped to record songs, take photographs, and make motion pictures of numerous other species they would encounter along the way. Their goal was not only to document these birds for science but also to make a profit through commercial distribution of the resulting motion pictures. Accompanying Allen on this fifteen-thousand-mile journey were Kellogg and another of his students, James T. Tanner. Albert Brand and George M. Sutton, a bird artist and curator of ornithology at Cornell, also joined the party at various points along the way.[23]

The Brand–Cornell University–American Museum of Natural History Ornithological Expedition set out from Ithaca in mid-February 1935 in two trucks, one filled with bulky equipment for sound recording, the other with a twenty-four-foot, collapsible observation tower mounted on its roof. The first destination was central Florida, south of Orlando. Here, on the banks of Taylor Creek, Allen had spotted a pair of ivory-billed woodpeckers during a 1924 visit, birds that a local taxidermist had subsequently collected and sold. Despite a monthlong search at that site and other areas of the state where the species had once been documented, Allen and his colleagues failed to locate a single living ivory-bill. Disappointed but undaunted, in late March the group headed to Madison Parish, Louisiana, where Pearson had enjoyed better luck three years previously. With the aid of J. J. Kuhn, a local woodsman and warden the Louisiana Department of Conservation had hired to enforce a ban on hunting at the Singer Tract, they renewed their search. Spring rains made the muddy roads impassible for their heavy trucks, so the naturalists loaded their supplies and hundreds of pounds of recording equipment onto a mule-drawn wagon and slogged their way into the heart of the Singer Tract, a thirty-mile-long by eighteen-mile-wide wilderness area bisected by the meandering Tensas River (see figure 6.1). On their third day at the site, seven miles from the nearest improved road, the expedition party was elated when Kuhn finally located a pair of nesting ivory-bills.[24] In the words of Sutton: "The whole experience was like a dream. There we sat in the wild swamp, miles and miles from any highway, with two ivory-billed woodpeckers so close to us that we could see their eyes, their long toes, even their strongly curved claws without binoculars."[25]

Figure 6.1

Transporting equipment into the Singer Tract, 1935. From *left* to *right*: James T. Tanner, J. J. Kuhn, Arthur A. Allen, "Albert," and "Ike." Kuhn, a local game warden who had intimate knowledge of the area's landscape, proved crucial to the success of Allen's expedition to photograph, film, and produce sound recordings of the endangered ivory-billed woodpecker at the Singer Tract. Later he would provide essential support for Tanner as he completed three seasons of fieldwork at that same remote, difficult-to-traverse site. *Photograph courtesy of James T. Tanner.*

Enchanted with their find, the group pitched a tent less than three hundred feet from the nesting site and dubbed their soggy new quarters "Camp Ephilus," a pun on the scientific name of the species Kuhn had located for them. Over the next several days they gathered a mountain of information on the life and behavior of the ivory-bill, much more data than had been previously accumulated on the species in the two centuries since Catesby's initial description. The naturalists took turns using tripod-mounted, twenty-four-power binoculars to monitor the nest cavity from dawn until dusk. In addition, Sutton sketched the birds from life, while Allen, Tanner, Kellogg, and Kuhn captured the first photographs, motion pictures, and sound recordings of the ivory-bill (see figure 6.2).

After remaining at the site for five days, members of the expedition party reluctantly broke camp. While they might have preferred to linger in the area to observe the young being reared, there were other species to film and record. Unfortunately, when Allen finally returned to the Singer Tract in early May, he

Figure 6.2

A pair of ivory-billed woodpeckers exchanging places on a nest, 1935. During his studies at the Singer Tract, where this photo was taken, James Tanner discovered that the species' restricted range was apparently related to its narrow feeding preferences. *Photograph courtesy of James T. Tanner.*

discovered that none of the newly hatched ivory-bills survived, either at the original site or at a second nesting location that Kuhn had located a mile to the north. Although clearly disappointed at the nesting failures of this critically endangered bird, Allen and his party were also delighted to have captured the sounds and images of the Lord God bird and more than one hundred fifty other species during their fifteen-thousand-mile journey.

Establishing the Audubon Fellowship

Allen's motion pictures and sound recordings featuring the elusive ivory-bill proved a hit at the numerous popular lectures he delivered following his return to Ithaca. Perhaps no audience was more enthralled with his presentation, however, than the members assembled for the National Association of

Audubon Societies annual meeting in October 1935.[26] The organization was devoting increasing attention to threatened species, in part because of an upheaval it had recently experienced. The sources of that turmoil were many, but much of it can be traced back to a scathing, sixteen-page pamphlet issued in 1929. Authored by two curators at the American Museum of Natural History in New York—Waldron D. Miller and Willard G. Van Name—and a young writer named Davis Quinn, *A Crisis in Conservation: Serious Danger of Extinction of Many North American Birds* rocked the organization to its very core.[27]

The primary target of the fiery pamphlet was the National Association of Audubon Societies, which T. Gilbert Pearson had been running for nearly two decades. On the surface the nation's largest wildlife conservation organization had flourished under Pearson's reign, with its membership rolls, annual budget, and endowment all steadily increasing. But according to critics, behind this rosy statistical facade the society had veered from its original course, becoming complacent and overly cozy with the government agencies and private organizations it should have been actively policing. While Audubon was standing by idly, Miller, Van Name, and Quinn charged, numerous birds were rushing headlong toward extinction, with scarcely a voice raised in protest. The great auk, Labrador duck, passenger pigeon, Carolina parakeet, and "also probably" the Eskimo curlew and the heath hen were already lost. A troubling number of additional species—the whooping crane, trumpeter swan, ivory-billed woodpecker, California condor, flamingo, golden plover, Hudsonian godwit, buff-breasted sandpiper, and upland plover—were also now "beyond saving." And as many as twenty-six additional species faced a similar fate in the near future. *A Crisis in Conservation* not only created much consternation within American wildlife conservation circles but also became the impetus for the creation of the Emergency Conservation Committee, a small organization that served as a gadfly in conservation circles and kept pressure on Audubon to reform its ways.[28]

It took several years for the dust to finally settle, but by 1934 Pearson had finally been forced from office, while John Baker, a bird watcher and no-nonsense business executive, took charge of Audubon's day-to-day operations. Over the next several years Baker instituted a series of reforms aimed at restoring the credibility and effectiveness of the organization.[29] One strategy for reinvigorating Audubon was to recruit new staff and board members. An example of the latter was Aldo Leopold, who had been making a name for himself through his thought-provoking conservation writings, his groundbreaking book *Game Management* (1933), and his newly created program in that same field at the University of Wisconsin, Madison, the first of its kind in the nation. Baker thought so highly of Leopold that at one point he praised him as one of the "most original thinkers on conservation and game management subjects I know."[30]

In the fall of 1935, Baker began discussions with Leopold on how to steer American wildlife conservation toward the needs of species facing extinction. Writing to Baker from Germany in September of that year, Leopold called for "an inventory of all native species, especially threatened ones, giving a pretty *national work plan* as to just what is needed to perpetuate that species, and who can do it. These items of needs should include not only protective legislation . . . but also protective actions other than legislation, and especially 'environmental control' needs. . . . Research needs should also be listed and 'assigned.'"[31] Leopold drew up a sample entry for the sandhill crane indicating exactly the kind of information he felt was critical for rescuing the species. Included was a call for a research fellowship to learn more about the life history of the bird.

Later that year Leopold began circulating a more fleshed-out version of this initial proposal, the draft of a paper titled "Proposal for a Conservation Inventory of Threatened Species."[32] There he pointed out that despite the "large and sudden increase" in wildlife conservation initiatives in the United States, governmental agencies, private organizations, and individual citizens had failed to coordinate effectively.[33] Moreover, too much activity remained focused on game species, which were likely to be saved by "powerful motives of local self-interest . . . however great the blunders, delays and confusion" in getting their management properly established. The same was not necessarily true for what Leopold referred to as "wilderness game" (like grizzly bears), migratory birds, predators, and rare plant associations, or "in general all wild native forms that fly at large or have only esthetic and scientific value to man." These species were the "threatened element in 'outdoor America'—the crux of conservation policy." What was desperately needed was a systematic listing of these threatened forms wherever they still survived, as well as an "inventory of the information, techniques, and devices applicable to each species in each place, and of local human agencies capable of applying them."[34] As an example of the kind of information he thought should be compiled, publicized, and acted on, Leopold pointed to the ivory-billed woodpecker, "a bird inextricably interwoven with our pioneer tradition—the very spirit of that 'dark and bloody ground' which has become the locus of national culture." Ornithologists had recently located a remnant population of these noble birds in a virgin Louisiana forest slated for logging. Now this discovery needed to be brought to the attention of relevant federal agencies—the National Park Service, the Forest Service, and the Bureau of Biological Survey—so they could coordinate to save this remnant habitat of the ivory-bill.[35]

About the time Leopold began circulating his proposal for an inventory of threatened species, Allen wrote him. While reviewing his notes on the ivory-billed woodpecker, the Cornell professor had become convinced of the need for a "well-qualified young ornithologist to make an intensive study of the present status and life history of the bird, taking up where we left off last spring."[36] Allen's student, James Tanner, had expressed interest in pursuing the project and even approached Audubon officials about the possibility of funding. A recent graduate from Cornell with honors in biology and ornithology, Tanner had proven a capable field ornithologist during the Brand–Cornell–American Museum Expedition, where he demonstrated "an ability to rough it and to get along with all kinds of people in all kinds of situations, a natural adaptability, ingenuity, initiative, originality, and willingness to work."[37] Leopold replied enthusiastically to the suggestion: "Nothing would delight me more than to see you follow up with a life history study of the ivory-bill. I would also consider it highly appropriate for the Audubon Society to finance such a study. . . . I hope, too, that any research plan would include an administrative plan for the perpetuation of the species."[38]

Baker, too, was quickly sold on the idea. However, he and his staff began to think in terms not simply of a single graduate fellowship to study the ivory-bill, but a series of Audubon-sponsored fellowships to investigate other endangered species as well. Besides discussions with Allen, Baker also corresponded with Joseph Grinnell, at the University of California, Berkeley, about a fellowship to study the California condor, and Charles T. Vorhies, at the University of Arizona, about another to study the desert mountain sheep. In his speech announcing the newly launched fellowship program at the 1936 Audubon meeting, Baker recommended that similar studies on the whooping crane, glossy ibis, sage grouse, grizzly bear, and the mountain lion should also be initiated.[39]

After unveiling the fellowship program, Baker wrote Allen and asked whether Cornell officials would support such an initiative, if Allen would be willing to recommend and supervise a suitable candidate, and when that individual could begin research. In Baker's view, the proposed study should consist of three parts: (1) a "thorough inventory" to discover how many and where ivory-bills still existed, (2) a "thorough ecological study of the effect of the environment on the bird," and (3) a detailed plan to preserve the remaining populations of the species. To ensure that the research fellow would conduct his studies with "impartiality" and an "unprejudiced approach," Baker thought Allen should choose the candidate. The results of the study would be issued in a widely circulated report that Audubon would publish. In subsequent correspondence Baker indicated that he was willing to commit up to $1,500 a year for three years to fund the Ivory-Bill Graduate Research Fellowship.[40]

Given the ongoing challenge of securing financing for his students, especially during the Depression, Allen was delighted with the Audubon proposal. During negotiations over the specifics of the contract, he tried to reassure Baker that, notwithstanding the fate of the recently lost heath hen, science might still rescue the ivory-bill:

> I realize that there are some who feel that when a species becomes rare as is the Ivory-bill that it is useless to try to preserve it, and they quote the history of the Heath Hen as an example. I feel, however, that had we known as much about Grouse in general twenty years ago as we do today, the Heath Hen might have been saved, and the same holds true for the Ivory-billed Woodpecker. Unless we know more than at present there is no hope of saving it. A thorough study may not bring forth the necessary facts for its preservation, but I am very hopeful that if it is at all possible we will discover the means.[41]

To raise an endowment to finance this and other Audubon fellowships, Baker issued a circular appeal featuring a full-color plate of a pair of ivory-bills Sutton had drawn. That fund-raising circular highlighted the dire straights the bird faced, its association with wilderness habitats, and the need to ground wildlife conservation efforts on secure scientific foundations:

> Few birds or mammals, anywhere in the world, are in as critically dangerous a situation as this species, so vividly portrayed in Dr. Sutton's painting. Only vigorous efforts on its behalf can save it. Unless active steps are taken at once, there is little hope that future generations will ever know the thrill of coming on this bird in its wilderness fastnesses. The problem of saving it is complicated by the fact that so little is known about its habitat and requirements. It is with the purpose of learning requisite facts upon which to build a conservation program that the fellowship has been established.[42]

Trailing the Ivory-Bill

Having secured three years of funding, Tanner set out on the trail of the ivory-bill in January 1937. He spent the first several months in the field trying to discover how many and where ivory-bills might still exist. Based on published records, rumors of sightings, specimens in collections, and examinations of maps, he identified about fifty potential sites to survey in seven southeastern states.[43] On two occasions Tanner also conducted aerial reconnaissance as a prelude to more exhaustive ground searches.

Information from local residents proved crucial to his search.[44] After quickly discovering that simply asking locals if they had ever seen an ivory-billed woodpecker almost invariably "elicited a blank stare," he developed a more elaborate procedure for eliciting information about the potential existence of the species in a given area. He began with a series of general questions about the area's wildlife and landscape to gauge how knowledgeable and reliable a particular informant seemed. He would then ask if the more common pileated woodpecker could be found in the region, using whatever local name by which the bird happened to be known, and whether two large woodpeckers inhabited the area. If the answer to the latter question proved "yes," he would then inquire about how his informant distinguished between the two birds. To facilitate that discussion, whenever he was in the field Tanner carried a pocket-sized drawing showing the ivory-billed and pileated woodpeckers side by side. Even when an interviewee claimed recent, firsthand experience with the ivory-bill, Tanner remained wary about concluding that the species still inhabited the area. Rather, "reports of natives gave a good indication of whether or not certain areas were worth further investigation."[45] Ironically, one of the most useful sources of information turned out to be the employees of the logging industry, who tended to have a detailed knowledge of local timber stands and the wildlife they contained.

When he identified what appeared to be suitable habitat in swampy forests, he searched high and low for ivory-bill holes and feeding signs: extensive scaling of bark on trees and large limbs that had recently died. He also listened carefully for the bird's distinctive call and the characteristic loud double rap of its beak, both of which he knew firsthand from his earlier encounter with the species in Louisiana. In addition to providing information on the potential existence of ivory-bills in a given area, local residents also often served as guides to help him navigate through the densely forested, muddy, and often disorienting habitat the species seemed to prefer. Although he completed most of his searches during the winter and spring, when the bird was active but before the dense leaf canopy restricted viewing opportunities, Tanner compared the enterprise to "searching for an animated needle in a haystack." After personally investigating fifty potential sites, he concluded that fewer than two dozen ivory-bill woodpeckers still remained in five areas: the Singer Tract in Louisiana; the Big Cypress Swamp, the Apalachicola River region, and the Gulf Hammock–Suwannee River regions in Florida; and possibly the Santee Swamp in South Carolina.[46]

In fact, he had seen with his own eyes only six of the mysterious birds, all of them on the Singer Tract in Madison Parish, Louisiana.[47] There, with the aid of his indispensable guide, J. J. Kuhn, he spent three seasons in the field—a total of over a year—documenting their ecology and life history. The two men worked closely together whenever Tanner visited the site, and Kuhn continued making observations after Tanner left.[48] During the days when Tanner was in residence, the two would often split up so they could cover more ground. At night, they returned to their camp at the field site, a cabin owned by the Singer Manufacturing Company. There, by the light of a kerosene lamp, each would recap the day's adventures. "We talked . . . just as Mark Twain's river pilots endlessly discussed the details of the river's course," Tanner later noted. "Although learning of the Ivory-bill and its life history was our goal, the forest was our working place and we had to know it—to find our way, to travel quickly, and to know where to hunt."[49] Tanner clearly had a tremendous amount of respect not only for the intimate knowledge Kuhn had acquired about the Singer Tract but also for his commitment to preserving the ivory-bill.

With Kuhn's ongoing aid, Tanner gradually pieced together the story of the ivory-bill. Based on historical records and site observations, he estimated that the maximum population density of the species was about one pair per six square miles, while the pileated woodpecker's density was six pairs per square mile, and the red-bellied woodpecker's twenty-one pairs per square mile.[50] The usual range of the ivory-bills he studied in spring (nesting) and summer seasons was about three-quarters to one mile, while the winter range proved more than twice that. Although the birds he observed in Louisiana seemed largely sedentary, the wide variability of bill lengths and body sizes in specimens collected from other locations suggested a lack of isolation in the species, while providing evidence that it "sometimes wandered considerable distances."[51]

Among his most important discoveries, Tanner found that the species' dire status was linked to its narrow feeding preferences. Although limited evidence suggested the bird would eat a variety of fruits, nuts, and seeds, its primary food source seemed to be the wood-boring larvae of several families of beetles that inhabit the bark and sapwood of recently dead tree trunks and branches. Ivory-bills were present only in forests where dead and dying trees were frequent and other woodpeckers abundant, Tanner argued, conditions normally found only in "tracts of uncut, mature timber."[52] There the birds energetically knocked off tree bark to obtain the beetle larvae they favored. The Singer Tract ivory-bills did the vast majority of their feeding on sweet gum, Nuttall's oak, and hackberries that ranged from twelve to thirty-six inches in diameter, while in Florida the bird usually seemed to locate its prey on cypress and pine.[53] The mature sweet gum–oak association seemed to be the preferred habitat of most ivory-bill populations outside Florida. But transient populations of the species sometimes appeared in other forest associations where large numbers of trees had recently been killed by storm, fire, or logging operations. They would then disappear within a few years, when those sites no longer attracted large numbers of the beetle larvae on which they fed. While overzealous collecting and careless hunting might have contributed to the decline of the species in some areas, Tanner concluded that the overall diminution in range of the ivory-bill correlated closely with the spread of the logging industry in the South. In order to save the species, remnants of those imposing forests must be spared from the axe: "Mature forests of large, old trees have almost disappeared, and these conditions favorable for the Ivory-bill will very probably never again prevail. Its preservation must be accomplished by saving suitable habitat or by maintaining on certain areas an adequate food supply for the birds."[54]

Based on his research findings, Tanner developed a set of policy recommendations that he hoped would rescue the species from the fate of extinction. Ideally, areas still inhabited by the ivory-bill should be preserved as "refuges and as primitive areas."[55] The most promising sites included the Big Cypress Swamp and the lower Apalachicola River swamps in Florida, possibly the Santee River swamp in South Carolina, and most important, his main study site, the Singer Tract in Louisiana. In these locations he recommended securing a minimum undisturbed area of about two and a half to three square miles for each pair of the birds. Tanner made it clear that his estimate represented a minimum requirement, and that "the larger the area of the forest the better for the conservation of the birds—that is certain."[56] Where the chief obstacle to the establishment of a large protective refuge was the market value of the standing timber, as with the Singer Tract, Tanner called for a program of selective cutting outside the undisturbed areas that would leave a sufficient number of dead and dying trees to supply borers to the woodpeckers. He also thought selective killing of suitable trees, to foster the growth of appropriate borer populations, might artificially increase the birds' food supply.[57]

Tanner's copiously illustrated account was published in 1942 as the inaugural volume of the Research Report of the National Audubon Society series.[58] In a foreword to the publication, Baker declared that Audubon considered scientific research of the type contained in this report to be an "essential basis for wise policies governing the conservation of wildlife resources," and he proclaimed that his organization was firmly committed to the proposition that "the Ivory-bill shall not, as a part of America's heritage, go the way of the Passenger Pigeon and the Great Auk."[59] Allen contributed a preface in which he declared that "the greatest tragedy in nature is the extinction of a species. . . . Where is the man who would knowingly stand by and watch a marvelous creation of nature—harmless to man's interests, and of no commercial value—be forced into the vortex of extirpation without raising his voice in protest? Surely no intelligent human being could be indifferent to the passing of the last Ivory-billed Woodpecker."[60] After alluding to the ominous clouds of war that were hanging over the United States, Allen appealed to his readers' sense of nationalism to save the species:

> Today we are measuring our love of freedom in billions of dollars and thousands of lives. The American way of living is worth anything we have to pay to preserve it, and the Ivory-billed Woodpecker is one little guide post on our way of life, a reminder of that pioneering spirit that has made us what we are, a people rich in resourcefulness and powerful to accomplish what is right. The Ivory-bill is a product of the great force of evolution acting on American bird life in ages past, to produce in our southeastern United States the noblest woodpecker of them all—one that inspired Mark Catesby and John James Audubon and Alexander Wilson—one that has lured scores of recent ornithologists to the cypress jungles of South Carolina, Florida, and Louisiana in the ardent hope of seeing one individual alive. Is it worth ten dollars to save it? Is it worth ten million dollars? It is worth whatever we must pay before it is too late.[61]

The Fate of the Singer Tract

Even before the publication of Tanner's final report, the National Audubon Society had begun efforts to protect the ivory-bill at several sites, but the organization focused most intensely on the Singer Tract. The area was named after the Singer Manufacturing Company, which had acquired the land in 1916. Ten years later company officials contracted with the Louisiana Department of Conservation to establish a game refuge in the area.[62] Their motivation was not to protect the ivory-bill or other wildlife in the area, however, but simply to minimize the risk of accidental fires started by careless hunters. By the time Tanner began his study in 1937, the Singer Tract consisted of roughly eighty-one thousand acres, about four-fifths of which he considered "virgin timber."[63] That same year the Chicago Mill and Lumber Company acquired rights to the lumber on the Singer Tract and soon began logging west of the Tensas River. By 1941 the tract was approximately 40 percent cutover.[64] Fortunately for the ivory-bill, most of the logging had taken place outside of areas where Tanner had found the species.

Baker pursued numerous avenues in a desperate bid to protect this prime ivory-bill habitat. He appealed directly to President Franklin D. Roosevelt for help, and soon officials in the Forest Service, National Park Service, and Fish and Wildlife Service became involved in discussions about preserving the Singer Tract.[65] He failed to convince any of these federal agencies to acquire the home of the ivory-bill, however, the primary obstacle being the value of the land's standing timber, estimated to be worth about two million dollars.[66]

Baker also pleaded with two successive governors of Louisiana to simply declare a wildlife sanctuary in the area using powers of condemnation, but both refused. With the pace of logging increasing during the war, in 1943 Baker obtained a two-hundred-thousand-dollar commitment from the state of Louisiana to purchase part of the remaining uncut area in the Singer Tract as an inviolate wildlife refuge. He also convinced the governors of Louisiana, Tennessee, Arkansas, and Mississippi to sign a petition requesting the Chicago Mill and Lumber Company to waive its contractual rights to cut the remaining timber on the Singer Tract and to sell a cutover buffer area for what seemed a reasonable price.[67] At a meeting in Chicago in December of that year, however, officials from Chicago Mill and Lumber flatly refused to negotiate. By that point German prisoners of war were logging the land and most of the lumber was being used to make shipping crates and tea chests for the British.[68] By 1944 only one ivory-bill remained in the Singer Tract. It was the last documented sighting of the bird in the area.[69] And still the cutting continued. (Ironically, nearly four decades later, in 1980, the federal government created the Tensas River Wildlife Refuge on part of the now largely reforested Singer Tract.)

Conclusion

In the end, hopes for saving the ivory-billed woodpecker were dashed on the hard rocks of political and economic reality. Clearly modern science could reveal crucial information about the status and ecology of this magnificent creature, but knowledge alone could not effect the fundamental reorientation in values needed to rescue the species from the brink of extinction. By the time the United States began to show signs of such a value shift in the late 1960s and early 1970s, it might have been too late for the ivory-bill. Although the species was officially listed soon after passage of the Endangered Species Preservation Act of 1966, by that point its trail seemed to have grown cold.[70] Occasional reports of sightings continued to nurse hopes that the bird might still grace our bottomland hardwood forests, but until quite recently it appeared that the ivory-bill remained only as lifeless skins in the drawers of museums, as fading images on paper and film, and as ghosts that haunt us with what might have been.[71]

Beyond a lack of political will to implement the recommendations that Tanner made, there were limitations in the science he pursued. His pioneering study clearly went far beyond previous research on endangered species, but it remained restricted in scope and scale. Tanner's Audubon-sponsored study had been undertaken with a modest budget, over a fairly narrow span of time, and by a single scientist working closely with a limited number of local residents. While Tanner traversed thousands of miles during the intensive phase of his study and scoured several locations where recent sightings of the species had been claimed, he managed to actually find and observe only six ivory-bills at a single site. Exactly how many of these charismatic birds still roamed the earth? Where did they live? How representative were the behaviors and feeding preferences Tanner and Kuhn witnessed at the Singer Tract? How much variability did the species exhibit in terms of its habitat needs? Despite Tanner's hard work, knowledge, and dedication, these critical questions remained unanswered following his study—and perhaps unanswerable for a single naturalist, no matter how capable and devoted.

Scientists routinely make universal knowledge claims based on data gathered in a highly local context. In Tanner's case, that attempt might have unintentionally misled conservationists to focus unduly on a particular habitat type—mature bottomland forests—as the natural home of the ivory-bill and on habitat destruction as the most immediate, dire threat the species faced. But as the conservation biologist Noel Snyder has recently argued, there is good reason to suspect that human predation—by subsistence hunters, bird collectors, and curiosity seekers—might have played a greater role in the decline of the ivory-billed woodpecker than habitat destruction. Indeed, as Snyder points out, at Tanner's primary study site, the Singer Tract, the number of ivory-bills had declined from fourteen to six between the years 1934 and 1938, yet timbering did not begin there until the very end of this period.[72]

The post–World War II period witnessed a dramatic increase in public support for endangered species preservation along with impressive growth in the size of American universities, trends that converged to generate more funding for research on the issue. At the same time, the passage of the Endangered Species Act of 1973—one of the strongest, most comprehensive pieces of environmental legislation ever enacted—presented a federal mandate for the study and preservation of organisms facing the threat of extinction.[73] Research that had once been completed by lone individuals and their guides, laboring in the field over limited periods, was increasingly undertaken by large teams whose activities extended over much longer time spans. Those teams struggled not only to disentangle the often complex causes of species decline, but also to develop increasingly interventionist and management-intensive responses to those declines.[74] By the mid-1980s, proponents of a new, mission-oriented, interdisciplinary enterprise dubbed "conservation biology" successfully forged a new scientific field that placed at its core the analysis and protection of the full range of the earth's biological diversity.[75]

One telling measure of the transformations in the decades since Tanner's pioneering study came in the winter of 2004, when Tim Gallagher, the editor in chief of the Cornell Lab of Ornithology's *Living Bird* magazine, and Bobby Harrison, an avid bird watcher and associate professor at Oakwood College in Huntsville, Alabama, traveled to the Cache River National Wildlife Refuge in Arkansas's Mississippi Delta.[76] What brought them to that remote swampy site was a recent claim of an ivory-bill sighting. On the second day of their search, a large, black-and-white bird that Gallagher and Harrison immediately recognized as an ivory-billed woodpecker darted out of sight less than one hundred feet in front of their canoe. Thrilled with their fleeting glimpse of the species that most experts considered extinct, Gallagher informed John Fitzpatrick, the director of the Cornell Lab of Ornithology, who quickly organized a large-scale search of the region involving not only Cornell staff, but also the Fish and Wildlife Service, the Nature Conservancy, and state wildlife officials. The initial search by a team of about a dozen field biologists produced several additional sightings and one grainy, four-second video of what was purported to be the long-lost ivory-bill. Between December 3, 2005, and April 30, 2005, eight full-time searchers and thirty part-timers expanded the search to the White River National Wildlife Refuge and installed a series of autonomous recording units in the hope of capturing the species' kent call or its characteristic double rap. In late April 2005, Fitzpatrick and his collaborators stunned the world when they called a widely covered press conference to declare that the ivory-bill still lived. During the next season, the search team grew to 22 full-time biologists and 112 volunteers, with a budget of over one million dollars. In conjunction with the Nature Conservancy, the National Audubon Society, and other conservation organizations, ivory-bill enthusiasts also raised funds for the purchase of thousands of acres of potential ivory-bill habitat in the region. Several highly placed skeptics continue to question the evidence produced thus far, but the point is that what seemed like credible sightings of this elusive species generated a massive search for conclusive verification, an outpouring of public interest in the endeavor, and the funds for the acquisition of critical ivory-bill habitat. Clearly much had changed in the seventy years since Tanner first set out on the trail of the ivory-bill.

Notes

I would like to thank Jeremy Vetter for his helpful advice and hard work in bringing this anthology to press, David Tomblin for providing useful feedback, Nancy Tanner for sharing the wonderful photographs of the Cornell team making their way into the Singer Tract and of the ivory-bills themselves, and the University of Chicago Press for allowing me to use portions of this chapter that appear in my book *Nature's Ghosts: Confronting Extinction from the Age of Jefferson to the Age of Ecology*,

1. The basic story of Tanner's study is told in a series of recent studies, including Christopher Cokinos, *Hope Is the Thing with Feathers: A Personal Chronicle of*

Vanished Birds (New York: J. P. Tarcher/Putnam, 2000), 61–117; Jerome A. Jackson, *In Search of the Ivory-Billed Woodpecker* (Washington, D.C.: Smithsonian Books, 2004), 114–152; Phillip Hoose, *The Race to Save the Lord God Bird* (New York: Farrar, Straus and Giroux, 2004), 69–124; and Tim Gallagher, *The Grail Bird: Hot on the Trail of the Ivory-Billed Woodpecker* (Boston: Houghton Mifflin, 2005). On the fate of the ivory-billed woodpecker from an environmental history perspective, see Mikko Saikku, "'Home in the Big Forest': Decline of the Ivory-Billed Woodpecker and Its Habitat in the United States," in *Encountering the Past: Essays in Environmental History*, ed. Timo Myllyntaus and Mikko Saikku, rev. ed. (Athens: Ohio University Press, 2001), 94–140. Although my emphasis is different, I have relied on each of these publications in the essay that follows. A convenient bibliography of writing on the species is Ruth Rehfus, *American Ivory-Billed Woodpecker (Campephilus p. principalis)* (Washington, D.C.: U.S. Department of Interior, Departmental Library, 1969).

2. On the destruction of southern bottomland forests, see Michael Williams, *Americans and Their Forests* (Cambridge: Cambridge University Press, 1989); and Mikko Saikku, *This Delta, This Land: An Environmental History of the Yazoo-Mississippi Floodplain* (Athens: University of Georgia Press, 2005).
3. The history of American wildlife management is treated in Thomas Dunlap, *Saving America's Wildlife* (Princeton: Princeton University Press, 1988); Curt Meine, *Aldo Leopold: His Life and Work* (Madison: University of Wisconsin Press, 1988); Kurkpatrick Dorsey, *The Dawn of Conservation Diplomacy: U.S. Canadian Wildlife Protection Treaties in the Progressive Era* (Seattle: University of Washington Press, 1998); Christian C. Young, *In the Absence of Predators: Conservation and Controversy on the Kaibab Plateau* (Lincoln: University of Nebraska Press, 2002); Aldo Leopold, *Game Management* (New York: C. Scribner's Sons, 1933); the series of articles in the special, fiftieth-anniversary edition of *Wildlife Society Bulletin* 15 (1987); Albert G. Way, "Burned to Be Wild: Herbert Stoddard and the Roots of Ecological Conservation in the Southern Longleaf Pine Forest," *Environmental History* 11 (2006): 500–526; Keir Sterling, "Zoological Research, Wildlife Management, and the Federal Government," in *Forest and Wildlife Science in America*, ed. Harold K. Steen (Durham, N.C.: Forest History Society, 1999), 19–65; and Bonnie Christensen, "From Divine Nature to Umbrella Species: The Development of Wildlife Sciences in the United States," in Steen, *Forest and Wildlife Science in America*, 202–229. On wildlife conservation in the United States during the late nineteenth and early twentieth centuries, see Lisa Mighetto, *Wild Animals and American Environmental Ethics* (Tucson: University of Arizona Press, 1991). Wildlife conservation in the national parks is treated in Richard Sellars, *Preserving Nature in the National Parks: A History* (New Haven: Yale University Press, 1997); Alfred Runte, *Yosemite: The Embattled Wilderness* (Lincoln: University of Nebraska Press, 1990); James A. Pritchard, *Preserving Yellowstone's Natural Conditions: Science and the Perception of Nature* (Lincoln: University of Nebraska Press, 1999); and R. Gerald Wright, *Wildlife Research and Management in the National Parks* (Urbana: University of Illinois Press, 1992). On the role of sport hunters in American wildlife conservation, see James A. Tober, *Who Owns the Wildlife? The Political Economy of Conservation in Nineteenth-Century America* (Westport, Conn.: Greenwood Press, 1981); John F. Reiger, *American Sportsmen*

and the Origins of Conservation (New York: Winchester Press, 1975); and James B. Trefethen, *An American Crusade for Wildlife* (New York: Winchester Press, 1975). The history of the Audubon movement is detailed in Frank Graham Jr., *The Audubon Ark: A History of the National Audubon Society* (New York: Knopf, 1990); Robin Doughty, *Feather Fashions and Bird Preservation: A Study in Nature Protection* (Berkeley and Los Angeles: University of California Press, 1975); and Mark V. Barrow Jr., *A Passion for Birds: American Ornithology After Audubon* (Princeton: Princeton University Press, 1998). Legal dimensions are stressed in Michael J. Bean, *The Evolution of National Wildlife Law*, rev. ed. (New York: Praeger, 1983), 89–98; and Thomas A. Lund, *American Wildlife Law* (Berkeley and Los Angeles: University of California Press, 1980).

4. The history of natural history and the field sciences more broadly are presented in Paul Farber, *Finding Order in Nature: The Naturalist Tradition from Linnaeus to E. O. Wilson* (Baltimore: Johns Hopkins University Press, 2000); N. Jardine, J. A. Secord, and E. C. Spary, eds., *Cultures of Natural History* (Cambridge: Cambridge University Press, 1996); Henrika Kuklick and Robert E. Kohler, eds., *Science in the Field, Osiris* 11 (1996); and Robert E. Kohler, *All Creatures; Naturalists, Collectors, and Biodiversity, 1850–1950* (Princeton: Princeton University Press, 2006). On the elevation of the doctoral degree as the standard entry-level degree in science, see Kohler, "The Ph.D. Machine: Building on the Collegiate Base," *Isis* 81 (1990): 638–662. On the transformation of American natural history and the development of biology, see Ronald Rainger, Keith R. Benson, and Jane Maienschein, ed., *The American Development of Biology* (Philadelphia: University of Pennsylvania Press, 1988); Keith R. Benson, Jane Maienschein, and Ronald Rainger, eds., *The Expansion of American Biology* (New Brunswick: Rutgers University Press, 1991); and Philip J. Pauly, *Biologists and the Promise of American Life* (Princeton: Princeton University Press, 2000). For an insider's account of the growth of the Cornell program, see Albert H. Wright, "Biology at Cornell University," *Bios* 24 (1953): 122–145; on the Berkeley program, see Richard M. Eakin, "History of Biology at the University of California, Berkeley," *Bios* 27 (1956): 67–80; and Eakin, *History of Zoology at Berkeley* (Berkeley and Los Angeles: University of California, 1988); and on the Michigan program, see A. Franklin Shull, "The Department of Zoology," in *The University of Michigan: An Encyclopedic Survey*, ed. Wilfred B. Shaw (Ann Arbor: University of Michigan Press, 1951), 2:738–750.
5. The story of the great auk is told in Errol Fuller, *The Great Auk* (New York: Henry N. Abrams, 1999); and Jeremy Gaskell, *Who Killed the Great Auk?* (New York: Oxford University Press, 2001). The story of the passenger pigeon is recounted in Jennifer Price, *Flight Maps: Adventures with Nature in Modern America* (New York: Basic Books, 1999), 1–55; and A. W. Schorger, *The Passenger Pigeon: Its Natural History and Extinction* (Madison: University of Wisconsin Press, 1955). These and other avian extinctions are also documented in Cokinos, *Hope Is the Thing with Feathers*; and James Greenway, *Extinct and Vanishing Birds of the World*, 2nd ed. (New York: Dover Publications, 1967).
6. Cokinos, *Hope Is the Thing with Feathers*, 121–193.
7. See, for example, Noel F. R. Snyder, *The Carolina Parakeet: Glimpses of a Vanished Bird* (Princeton: Princeton University Press, 2004); Mikko Saikku, "The Extinction of the Carolina Parakeet," *Environmental History Review* 14 (1990): 1–18; Cokinos,

Hope Is the Thing with Feathers, 7–58; Daniel McKinley, *The Carolina Parakeet in Florida* (Gainesville: Florida Ornithological Society, 1985); Noel F. R. Snyder and Helen Snyder, *The California Condor: A Saga of Natural History and Conservation* (New York: Academic Press, 2000); John Nielsen, *Condor: To the Brink and Back—The Life and Times of One Giant Bird* (New York: HarperCollins, 2006); Robin W. Doughty, *Return of the Whooping Crane* (Austin: University of Texas Press, 1985); Thomas R. Dunlap, "Organization and Wildlife Preservation: The Case of the Whooping Crane in North America," *Social Studies of Science* 12 (1991): 197–221; and Winston E. Banko, *The Trumpeter Swan: Its History, Habits, and Population in the United States* (1960; repr., Lincoln: University of Nebraska Press, 1980).

8. The term "indigenous knowledge" refers to the "unique, traditional, local knowledge existing within and developed around the specific conditions of women and men indigenous to a particular geographic area." Louise Grenier, *Working with Indigenous Knowledge: A Guide for Researchers* (Ottawa: International Development Research Centre, 1998), 1. It is often used interchangeably with the terms "local" and "traditional" knowledge. These terms first began to appear in discussions around sustainable development, natural resource management, and intellectual property rights in non-Western nations during the 1980s, but they did not really take off until the 1990s. Here I am using the term "indigenous knowledge" more loosely to describe the intimate, firsthand knowledge about nature that local residents in *any* given community, particularly a rural community, acquire through their daily interactions with the natural world. My thinking about the importance of indigenous knowledge has also been influenced by the insights of Richard White, who stresses how individuals come to know the natural world through their physical labor on particular landscapes. See his article "'Are You an Environmentalist or Do You Work for a Living?': Work and Nature," in *Uncommon Ground: Toward Reinventing Nature*, ed. William Cronon (New York: Norton, 1995), 171–185; and his book *The Organic Machine: The Remaking of the Columbia River* (New York: Hill and Wang, 1995). See also David Turnbull, *Masons, Tricksters, and Cartographers: Comparative Studies in the Sociology of Scientific and Indigenous Knowledge* (Amsterdam: Harwood Academic, 2000), which argues that much of modern science is actually a form of indigenous or local knowledge; Daniel W. Schneider, "Local Knowledge, Environmental Politics, and the Founding of Ecology in the United States: Stephen Forbes and 'The Lakes as a Microcosm,'" *Isis* 91 (2000): 681–705, which offers an insightful case study of how local knowledge played a key role in the birth of ecology; and Robert E. Kohler, *All Creatures: Naturalists, Collectors, and Biodiversity, 1850–1950* (Princeton: Princeton University Press, 2006), 14, which makes a similar distinction between "cosmopolitan" and "residential" knowledge, while arguing that the best field naturalists manage to combine elements of both. For a more skeptical view of the distinction between indigenous and expert knowledge, see Arun Agrawal, "Indigenous and Scientific Knowledge: Some Critical Comments," *Indigenous Knowledge and Development Monitor* 3, no. 3 (2000): 3–9.

9. Catesby's engraving of the ivory-bill and the accompanying text can be found in a convenient modern edition of his *Natural History of Carolina, Florida, and the Bahaman Islands* (1731–1743): Alan Feduccia, ed., *Catesby's Birds of Colonial America* (Chapel Hill: University of North Carolina Press, 1985), plate 1, 88–89.

10. Alexander Wilson, *Wilson's American Ornithology* (1808–1814; repr., Boston: Otis, Broaders, 1839), 275.
11. Scott Russell Sanders, ed., *Audubon Reader: The Best Writings of John James Audubon* (Bloomington: Indiana University Press, 1986).
12. On the culture and practice of bird and egg collecting, see Barrow, *Passion for Birds*; Barbara Mearns and Richard Mearns, *The Bird Collectors* (San Diego: Academic Press, 1998); and Carol R. Henderson, *Oology and Ralph's Talking Eggs: Bird Conservation Comes Out of Its Shell* (Austin: University of Texas Press, 2007).
13. The key role of Chapman and Brewster in the Audubon movement is documented in Graham, *Audubon Ark*; and Barrow, *Passion for Birds*.
14. See Elizabeth Austin, ed., *Frank M. Chapman in Florida: His Journals and Letters* (Gainesville: University Press of Florida, 1967); the correspondence between Chapman and Brewster from this period in the Brewster Papers, Museum of Comparative Zoology, Harvard University, Cambridge, Mass.; and William Brewster and Frank M. Chapman, "Notes on the Birds of the Lower Suwanee River," *Auk* 8 (1891): 136–137.
15. Paul Hahn, *Where Is That Vanished Bird?* (Toronto: University of Toronto Press, 1963) lists the locations of over four hundred specimens. Jerome Jackson reported twenty plus additional specimens in "The Ivory-Billed Woodpecker," in *Rare and Endangered Biota of Florida*, vol. 5, *Birds*, ed. James A. Rodgers Jr., Herbert W. Kale II, and Henry T. Smith (Gainesville: University Press of Florida, 1996), 103–112. Other published accounts of collectors in search of the ivory-bill include Frederic H. Kennard, "On the Trail of the Ivory-Bill," *Auk* 32 (1915): 1–14; and Robert Ridgway, "The Home of the Ivory-Bill," *Osprey* 3 (1898): 35–36.
16. T. Gilbert Pearson, "Protection of the Ivory-Billed Woodpecker," *Bird-Lore* 34 (1932): 300–301.
17. Reported in Cokinos, *Hope Is the Thing with Feathers*, 67. Biographical information on Allen may be found in Olin S. Pettingill Jr., "In Memoriam: Arthur A. Allen," *Auk* 85 (1968): 192–202; Edwin Way Teale, "Arthur A. Allen: Ten Thousand Bird Students Have Learned from Him," *Audubon Magazine* 45 (1943): 84–89; and Richard B. Fisher, "Ambassador of Birdlife," *Audubon Magazine* 67 (1965): 26–31.
18. The history of Cornell program is discussed in Arthur A. Allen, "Ornithological Education in America," in *Fifty Years' Progress of American Ornithology, 1883–1933*, ed. Frank. M. Chapman and T. S. Palmer (Lancaster, Pa.: American Ornithologists' Union, 1933), 215–229; Allen, "Cornell's Laboratory of Ornithology," *Living Bird* 1 (1962): 7–36; and Gregory S. Butcher and Kevin McGowan, "History of Ornithology at Cornell University," *Contributions to the History of North American Ornithology*, ed. William E. Davis and Jerome A. Jackson (Cambridge, Mass.: Nuttall Ornithological Club, 1995), 223–245.
19. On the nature study movement, see Sally Gregory Kohlstedt, "Nature, Not Books: Scientists and the Origins of the Nature-Study Movement in the 1890s," *Isis* 96 (2005): 324–352; and Kevin Armitage, *The Nature Study Movement: The Forgotten Popularizer of America's Conservation Ethic* (Lawrence: University Press of Kansas, 2009).
20. On recording birdsong at the Cornell program, see Butcher and McGowan, "History of Ornithology," 233–34; Peter Paul Kellogg, "Bird-Sound Studies at Cornell," *Living Bird* 1 (1962): 37–48; and Jeffery Boswell and Dominic Couzens,

"Fifty Years of Bird Sound Publication in North America: 1931–1981," *American Birds* 36 (1982): 924–943.

21. See the extensive correspondence between Brand and Allen related to the planning of the expedition in Arthur A. Allen Papers, Rare and Manuscripts Collections, Carl A. Kroch Library, Cornell University, Ithaca, N.Y. (hereafter cited as Allen Papers, Cornell), Box 52.
22. Quoted from Allen's published account of the expedition: Arthur A. Allen, "Hunting with a Microphone the Voices of Vanishing Birds," *National Geographic* 71 (1937): 696–723, on p. 699. See also Arthur A. Allen and P. Paul Kellogg, "Recent Observations on the Ivory-Billed Woodpecker," *Auk* 54 (1937): 164–184; and Arthur A. Allen, "Ivory-Billed Woodpecker," in *Life Histories of North American Woodpeckers*, ed. Arthur C. Bent, Bulletin of the United States National Museum 174 (Washington, D.C.: Government Printing Office, 1939).
23. On Sutton's life and work, see Jerome A. Jackson, *George Miksch Sutton: Artist, Scientist, Teacher* (Norman: University of Oklahoma Press, 2007).
24. Arthur Allen to Louis Boochever, 15 April 1935, Allen Papers, Cornell, Box 55, gives full credit for the rediscovery of the ivory-bill to J. J. Kuhn.
25. George Miksch Sutton, *Birds in the Wilderness: Adventures of an Ornithologist* (New York: Macmillan, 1936), 195.
26. See the glowing report in "Audubon Association Holds Annual Meeting," *Bird-Lore* 37 (1935): 429–435, on p. 431.
27. Waldron DeWitt Miller, Willard G. Van Name, and Davis Quinn, *A Crisis in Conservation: Serious Danger of Extinction of Many North American Birds* (New York: n.p., 1929). For the circumstances leading up to this publication, see Barrow, *Passion for Birds*, 146–150; and Barrow, "Science, Sentiment, and the Specter of Extinction: Reconsidering Birds of Prey During America's Interwar Years," *Environmental History* 7 (2002): 69–98.
28. More details on the chain of events that led to the creation of the ECC and the organization's impact on American wildlife conservation are found in the recent biography of one of its cofounders, Dyana Z. Furmansky, *Rosalie Edge, Hawk of Mercy: The Activist Who Saved Nature from the Conservationists* (Athens: University of Georgia Press, 2009). See also Irving Brant, *Adventures in Conservation with Franklin D. Roosevelt* (Flagstaff, Ariz.: Northland, 1988); and Stephen Fox, *The American Conservation Movement: John Muir and His Legacy* (Madison: University of Wisconsin Press, 1985), 173–182.
29. On reforms in the National Association of Audubon Societies during this period, see Fox, *American Conservation Movement*, 173–182; and Graham, *Audubon Ark*, 107–144.
30. John Baker to Guy Emerson, 27 November 1940, National Audubon Society Records, Manuscripts and Archives Division, New York Public Library, New York (hereafter cited as NAS Records, NYPL), Box B-4, Folder: Guy Emerson. Leopold was invited on to the board of the National Association of Audubon Societies in 1936.
31. Aldo Leopold to John Baker, 30 September 1935, NAS Records, NYPL, Box B-5, Folder: Ideas File.
32. See, for example, the copy in Box B-9, Folder: Speeches by Others, NAS Records, NYPL. Other copies may be found in Box 58, Allen Papers, Cornell; and Folder: Leopold, Museum of Vertebrate Zoology Records, University of California, Berkeley. Leopold published a version of this paper as "Threatened Species: A

Proposal to the Wildlife Conference for an Inventory of the Needs of Near-Extinct Birds and Animals," *American Forests*, March 1936, 116–119. This paper has also been reprinted in Aldo Leopold, *The River of the Mother of God, and Other Essays by Aldo Leopold* (Madison: University of Wisconsin Press, 1991), 230–234.

33. Leopold, "Proposal," 1.
34. Ibid., 2.
35. Ibid., 2–3.
36. Arthur A. Allen to Aldo Leopold, 21 April 1936, Allen Papers, Cornell, Box 58.
37. James T. Tanner, *Ivory-Billed Woodpecker*, Research Report of the National Audubon Society 1 (New York: National Audubon Society, 1942), iii.
38. Aldo Leopold to Arthur A. Allen, 28 April 1936, Allen Papers, Cornell, Box 58. See also Allen to Leopold, 4 May 1936; Leopold to Allen, 13 May 1936; and Allen to Leopold, 25 May 1936, for additional discussion. Allen was hoping that the Technical Committee of the recently created American Wildlife Institute, an organization that Leopold headed, might provide funds for the fellowship, but Baker rejected this idea. In the hope of finding funding, Allen helped Tanner craft a two-page proposal that was sent to Leopold and Baker. A copy of this proposal may be found in this same series of letters.
39. Much of this speech is reproduced in "The Audubon Research Fellowship Plan," *Bird-Lore* 38 (1936): 444–446. Additional information on the fellowship plan may be found in "Action on Threatened Species," *Bird-Lore* 39 (1937): 20–24. See also Richard Pough, "An Inventory of Threatened and Vanishing Species," *Transactions of the Second North American Wildlife Conference* (1937): 599–604.
40. See the long series of exchanges regarding the fellowship: John Baker to Arthur A. Allen, 18 November 1936; Allen to Baker, 25 November 1936; Baker to Allen, 4 December 1936; Baker to Allen, 9 December 1936; Baker to Allen, 10 December 1936; Baker to Allen, 15 December 1936; Allen to Baker, 17 December 1936; Baker to Allen, 23 December 1936; Allen to Baker, 28 December 1936; and "Special Temporary Fellowship Memorandum of Understanding," Allen Papers, Cornell, Box 57.
41. Allen to Baker, 17 December 1936, Allen Papers, Cornell, Box 57.
42. National Association of Audubon Societies, *Help Save the Ivory-Billed Woodpecker* (n.p., n.d.), located in Box B-10, Folder: Staff Memos, NAS Records, NYPL.
43. This phase of the study is discussed in Tanner's final report, *Ivory-Billed Woodpecker*, 20–29.
44. See, for example, Allen to Baker, 28 December 1938, Allen Papers, Cornell, Box 57.
45. Tanner, *Ivory-Billed Woodpecker*, 20–21.
46. Ibid., 20, 99.
47. In 1937, the first year of his study, Tanner personally observed two ivory-bill pairs, a lone male, and a young bird on the Singer Tract, but he accepted as credible claims by Kuhn and others that seven other ivory-bills resided there. Ibid., 39.
48. Indeed, Kuhn had begun systematically observing the ivory-bills on the Singer Tract as early as 1936. See J. J. Kuhn to Arthur A. Allen, 21 April 1936; and Allen to Kuhn, 28 April 1936, Allen Papers, Cornell, Box 58.
49. Quoted in Hoose, *Race to Save the Lord God Bird*, 105.
50. Tanner, *Ivory-Billed Woodpecker*, 33
51. Ibid., 36.
52. Ibid., 99.

53. Ibid., 42.
54. Ibid. 100.
55. Ibid.
56. Ibid., 56.
57. Ibid., 94–97.
58. Tanner's conclusions were first published in his dissertation, "The Life History and Ecology of the Ivory-Bill" (Ph.D. diss., Cornell University, 1940). There were four other publications in this Research Reports of the National Audubon Society series: Robert P. Allen, *The Roseate Spoonbill* (New York: National Audubon Society, 1942); Carl B. Koford, *The California Condor* (New York: National Audubon Society, 1953); Robert P. Allen, *The Whooping Crane* (New York: National Audubon Society, 1952); and Robert P. Allen, *The Flamingos: Their Life History and Survival, with Special Reference to the American or West Indian Flamingo (Phoenicopterus ruber)* (New York: National Audubon Society, 1956). Tanner also published a popular account of his studies as "Three Years with the Ivory-Billed Woodpecker, America's Rarest Bird," *Audubon Magazine* 43 (1941): 5–14.
59. Baker, foreword to Tanner, *Ivory-Billed Woodpecker*, ii.
60. Allen, preface to Tanner, *Ivory-Billed Woodpecker*, iii.
61. Ibid., iv.
62. In addition to the history in Tanner's report (37–39), see also Richard H. Pough, "Report to the Executive Director, National Audubon Society, on the Present Condition of the Tensas River Forests of Madison Parish, Louisiana, and the Status of the Ivory-Billed Woodpecker in This Area as of January, 1944," MS, NAS Records, NYPL, Box B-9, Folder: Singer Tract, 1944.
63. Tanner, *Ivory-Billed Woodpecker*, 37.
64. Ibid., 90.
65. See John H. Baker, "Statement with Regard to Establishment of Wildlife Refuge in Louisiana in an Effort, Among Other Things, to Preserve America's Rarest Bird," MS of speech delivered at Convention of Outdoor Writers Association of America in Columbus, Ohio, February 22, 1944, NAS Records, NYPL, Box B-9, Folder: Singer Tract, 1944; and Cokinos, *Hope Is the Thing with Feathers*, 101.
66. See the estimate in V. H. Sonderegger, "Inspection Report on the Singer Reserve in Madison Parish, Louisiana, March 31, 1933," NAS Records, NYPL, Box B-9, Folder: Singer Tract, 1936–1943; and W. B. Bell, acting chief of the Bureau of Biological Survey, to Arthur A. Allen, 1 February 1937, Allen Papers, Cornell, Box 60.
67. From a copy of the petition in NAS Records, NYPL, Box B-9, Folder: Singer Tract, 1936–1943.
68. "For the Confidential Information of Directors of National Audubon Society, December 15, 1943," NAS Records, NYPL, Box B-9, Folder: Singer Tract, 1936–1943.
69. Reported in Don Eckleberry, "Search for the Rare Ivory-Bill," in *Discovery: Great Moments in the Lives of Outstanding Naturalists*, ed. John K. Terres (Philadelphia: J. B. Lippincott, 1961), 195–207.
70. The species was officially listed as endangered on March 11, 1967.
71. In 1986 the U.S. Fish and Wildlife Service appointed an Ivory-Billed Woodpecker Advisory Committee, consisting of Jerome Jackson, Lester Short, and James Tanner, to determine the status of the species. Between 1987 and 1989, Jackson received funding to systematically search promising areas of the Southeast for the

species. He did not see any ivory-bills, but in 1987 he did hear what he and his student thought had to be an ivory-bill just north of Vicksburg, Mississippi. He later concluded that "this effort might have been too little, too late." Jackson, "Ivory-Billed Woodpecker," 110. See also his *In Search of the Ivory-Billed Woodpecker.* There were several reliable sightings of ivory-bills in Cuba during the mid-1980s. See, for example, Lester L. Short and Jennifer F. M. Horne, "I Saw It!" *International Wildlife,* March–April 1987, 22–24; and Short and Horne, "The Ivory-Bill Still Lives," *Natural History,* July 1986, 26–28. Unfortunately, recent efforts to find the species in Cuba have not been successful. See Martjan Lammertink, "No More Hope for the Ivory-Billed Woodpecker," *Cotinga* 3 (1995): 45–47.

72. Noel F. R. Snyder, *An Alternative Hypothesis for the Cause of the Ivory-Billed Woodpecker's Decline* (Camarillo, Calif.: Western Foundation of Vertebrate Zoology, 2007); and Noel F. R. Snyder, David E. Brown, and Kevin B. Clark, *Travails of Two Woodpeckers: Ivory-Bills and Imperials* (Albuquerque: University of New Mexico Press, 2009). Both works note that at many sites, including the Singer Tract, the ivory-bill seemed to decline before there was substantial logging.
73. On the history, context, and consequences of the various Endangered Species Acts, see Shannon Petersen, *Acting for Endangered Species: The Statutory Ark* (Lawrence: University Press of Kansas, 2002); Steven Lewis Yaffee, *Prohibitive Policy: Implementing the Federal Endangered Species Act* (Cambridge: MIT Press, 1982); and Charles C. Mann and Mark L. Plummer, *Noah's Choice: The Future of Endangered Species* (New York: Knopf, 1995).
74. For example, in the case of the California condor, in 1985 the Fish and Wildlife Service issued a controversial emergency order calling for the removal of all remaining condors from the wild so that they could be incorporated into a captive breeding program designed to rescue the species. See Nielsen, *Condor*; and Snyder and Snyder, *California Condor.*
75. Curt Meine, Michael Soulé, and Reed F. Noss, "'A Mission-Driven Discipline': The Growth of Conservation Biology," *Conservation Biology* 20 (2006): 631–651; and Timothy J. Farnham, *Saving Nature's Legacy: Origins of the Idea of Biological Diversity* (New Haven: Yale University Press, 2007).
76. The story of the recent rediscovery of the ivory-billed woodpecker is told in Tim Gallagher, *The Grail Bird: Hot on the Trail of the Ivory-Billed Woodpecker* (Boston: Houghton Mifflin, 2005). The original claim of rediscovery was published as J. W. Fitzpatrick and others, "Ivory-Billed Woodpecker (*Campephilus principalis*) Persists in Continental North America," *Science* 308 (2005): 1460–1462. One of the more prominent skeptics of the recent claims is Jerome A. Jackson; see his "Ivory-Billed Woodpecker (*Campephilus principalis*): Hope, and the Interfaces of Science, Conservation, and Politics," *Auk* 123 (2006): 1–15. See also Cornell Laboratory of Ornithology Web site, "The Search for the Ivory-Billed Woodpecker," http://www.birds.cornell.edu/ivory/; and the blog maintained by Tom Nelson, "Ivory-Bill Skeptic," http://tomnelson.blogspot.com/. A team of scientists has made claims of a second rediscovery site in the panhandle of Florida; see Geoffrey E. Hill, *Ivory-Bill Hunters: The Search for Proof in a Flooded Wilderness* (New York: Oxford University Press, 2007). As of the time this manuscript went to press, these latest efforts have yet to produce definitive evidence—a clear photograph, motion picture, or specimen—that the bird still exists.

Seven

Helen M. Rozwadowski

Playing By—and On and Under—the Sea

The Importance of Play for Knowing the Ocean

The ocean has, for most of Western history, been perceived as uniquely separate from human civilization, untouchable and unchangeable, even more so than, for example, the North American wilderness before the arrival of Europeans. One of the main outcomes of environmental history has been to redress the cultural perception of "virgin wilderness," to explain how people have always changed nature, and to argue that people have been, and should be, considered part of nature. Scholars are just beginning to attempt the same for the ocean.

One of the problems with considering the ocean in relation to human history emerges from the observation that people do not live there. The ocean may be more like the cultural construct of "wilderness" than like other terrestrial space. It may share important qualities in common with places such as the polar regions, glaciers, remote mountains, the upper atmosphere, or outer space. Like all those places, people do not normally or comfortably live there. They do, however, work, play, fight, explore, and reflect at, and under, the ocean. Although most glimpses out to sea reveal endless waves reaching to the horizon rather than any lasting evidence of human presence, there are myriad human activities through which historical connections between people and oceans can be explored.[1]

In considering how to study the ocean historically, environmental historian Richard White's injunction to consider work as a critical category for knowing nature suggests a way forward.[2] We know the oceans through technologies and knowledge systems such as compasses, charts and navigational knowledge, or nets, bait, and knowledge of how to find fish. Historians could investigate the work of fishermen, navigators, sailors, civil engineers, tide-table makers, pearl divers, dredgers, seaweed collectors, and the innumerable other people who have made a living from or at sea. Their activities and experiences could provide access to historical understanding of the ocean in ways parallel to White's study of explorers, native salmon fishers, and dam builders along the Columbia River.[3]

Several aspects of White's blueprint for considering work are especially applicable to work as a category for knowing the ocean. First is his point that bodily knowledge of the natural world yields knowledge of nature: "Humans have matched their energy against the energy of flowing water and wind. They have known distance as more than abstraction because of the physical energy they expended moving through space."[4] The importance of bodily knowledge seems also to apply to the sea; think of the knowledge navigators had of water depth derived from handling sounding lines as they hauled them in. Here, experience mattered. Mariners without experience would not be as readily able to transform sensory perception into knowledge. In the case of sounding, the experienced navigator recognized the feel of the lead hitting bottom and was, therefore, able to get an accurate measurement of depth.

Although work appears to be a critical category for knowing the oceans, this essay argues that play is also indispensable. For the ocean, play is probably so inextricably connected with work that the two categories cannot always be separated. White speaks critically of what he perceives to be overemphasis on play, a category that terrestrial environmental historians have considered from early in that field's genesis, such as through attention to the creation of national parks. It is relevant to note that White's concerns lie with environmentalists (not professional historians), popular cultural perception, and the political danger of allowing opponents of environmentalism, through the wise use movement, an exclusive claim to work within nature.[5] "Environmentalists," he diagnosed, "stress the eye over the hand, the contemplative over the active, the supposedly undisturbed over the connected."[6] White's criticism might suggest that play, as a frivolous category, should also be shunned as a means for knowing nature. His own scholarship and that of other environmental historians such as William Cronon have attended to work, and have made clear that work as a historical category related to nature applies to native residents of North American and colonizing Europeans, and to labor employing both traditional and modern technology.[7] If environmentalists ought to heed White's message to eschew play in favor of work, the question remains whether this advice applies also to historians. For the ocean, the answer must surely be no.

White points out that all people work. Play is likewise a deeply human activity. It can involve bodily knowledge of nature, such as when kayakers and canoers know distance along a river by matching their energy against its flow. White does not deny that recreation conveys bodily knowledge, but he claims that play matters less than work: "There is nothing essential lost when recreation is broken off or forgone."[8] Play, however, has become work—or at least has become indistinguishable from work—in several senses. First, tourism and recreation have in the modern era grown to the status of major industries. While environmentalists who sign up for ecotours might be knowing nature in the terms White criticizes, the guides, agents, and others who work in the infrastructure of these industries undertake the same activities as part of their work. Their knowledge of nature seems very similar to that of loggers, fishermen, and other blue-collar workers. Indeed, they may have shifted from the one kind of work to the other during their careers. Nor is this transition limited in time to the present. In coastal communities throughout Britain and France in the late eighteenth through nineteenth centuries, watermen turned from fishing and other kinds of work at sea and along the coasts to rowing tourists around, taking them out on fishing expeditions, serving as their bathing assistants, and so on.[9]

Historian of science Robert Kohler explicates a second way, less obvious and equally if not more important, in which play became work. For middle-class, white-collar workers in the decades just before and after the turn of the twentieth century, recreation was pursued with a seriousness of purpose, to distinguish it from idleness. Leisure acquired moral qualities through the transformation of recreation into "an essential complement to the work of shop or office, a kind of work."[10] Kohler chronicles the development of twentieth-century biological surveys through which scientists aimed to gather a representative sample of fauna and flora through which they could study biogeography, speciation, and other pressing questions of the turn-of-the-century biological sciences. Whereas earlier collectors had sought novel specimens, a new generation of trained, professional scientists, usually working in museums, took to the field themselves to make collections based on their expert scientific knowledge. Their fieldwork resembled nothing so much as camping, hunting, fishing, and other outdoor activities that had recently become popular and acceptable as activities for respectable, middle-class people.

Hunting, for instance, a pastime previously associated with either leisured aristocrats or the poaching poor, became acceptable to middle-class people as a recreational activity precisely because it came to be perceived as honest, outdoor work of the sort that would provide white-collar workers a healthy and respectable outlet. Indeed, "science played a similar role [as recreational walking had done] in the cultural transformation of hunting and fishing into respectable middle-class recreations."[11] This was because hunting and fishing were remarkably similar in practice to natural history collecting. The work of science and the play of vacationing formed a feedback loop through which people were drawn to the inner frontiers of the United States, where their experiences fostered conceptions of nature that, in turn, reinforced the new social habits of outdoor enjoyment. For the scientists themselves, fieldwork and play were indistinguishable; many curators collected during their own vacations, while museum directors viewed somewhat suspiciously paid time that curators spent away from their desks. The knowledge these field scientists created of North American biodiversity was a direct product of their work-play.[12]

Kohler adds a valuable dimension to White's analysis through his focus on the middle classes. In invoking work, White begins by discussing logging, farming, and other productive, extractive, blue-collar work, then shifts to white-collar work such as his, in which typing at his computer nevertheless involves a connection to nature. White's reminder that all work connects workers to nature is well taken. He addresses this reminder to environmentalists, to convince them that even work that appears to tread lightly on the environment changes nature. He does not, in his short essay, explore the social history behind the two very different kinds of work represented by the categories of logging and writing on a computer. Kohler's history of scientists, who are a subset of the middle class, usefully pursues connections between work—and play—for one particular group of white-collar workers. In his case, curators harvested resources from nature (specimens), but otherwise their work more closely resembled that of office workers than loggers. If recreation turns out to be such an essential category for this one group of middle-class workers, then perhaps it might also be for others whose work similarly combines elements of work and play, such as, perhaps, modern sailors of traditionally rigged vessels or workers in the ecotourism industry. Kohler's scholarship suggests that much of what White says about work can, in certain instances, apply to play.

What of White's declaration that work has more consequences for nature than play? Work, he reminds us, has changed nature. When the logger is finished working, trees are dead.[13] But play can have equally serious consequences for nature, both positive and negative. Skin diving and spearfishing denuded undersea areas of abalone and fish, particularly after the widespread adoption of scuba technology by recreational divers. Much earlier, sport hunting decimated waterfowl populations in wetlands areas. Some hunting enthusiasts, witnessing the destruction caused by their sport, became involved in conservation efforts. Automobile tourists laid waste to parks such as Grand Canyon. Since the early nineteenth-century start of recreational yachting, boating has prompted the construction of piers and marinas and led to marine pollution from paints, coatings, and waste disposal. The popularity of beach holidays led to the reshaping of coasts, with boardwalks, groins, break walls, and sand dredging. Kohler addresses this issue, saying, "Mass vacationing physically transformed the environment of tourist zones to fit conceptions of 'nature' and 'wilderness' as surely as commercial agriculture turned savannas and prairies into corn and wheat belts."[14]

Both work and play, then, changed nature. White makes a second point that is critical for the project of knowing the ocean. He insists that labor with modern machines be included in the category of work that led to knowledge of nature. Dams and dam building, for example, were technologies and means through which humans have known the Columbia River through labor. This is a crucial point for history related to the sea, because people have mostly known the ocean indirectly, through the mediation of technologies such as fishing lines, sounding gear, hulls, sails, diving helmets, scuba gear, submarines, charts, and tide tables.[15]

Technology is no less a critical category for play than for work. Scuba equipment, for example, has had both constructive and destructive consequences for the oceans. Other forms of play, both terrestrial and marine, likewise involved technology, such as ski lifts and snowmaking, cars to transport campers to parks or sailors to yacht clubs, and such equipment as nylon lines, Kevlar sails, and neoprene wet suits. For both this point and White's point about bodily knowledge, it seems, analytical categories that apply to rivers also apply to the sea.

When historians think of oceans, they may think of naval battles, voyages of exploration, imperialism, global shipping, whaling, commercial fishing—in short, of serious work. Most of us would think first of beaches, swimming, surfing, sunbathing, boating, fishing, and other recreational pastimes. We have historically known the ocean through work.[16] Yet, for the modern era, people have also known the ocean through play. Indeed, people have seriously studied the ocean through play, including through marine biology, some branches of oceanography, underwater archaeology, and also the sailing of traditionally rigged vessels. And play has also contributed motive and context for the development of knowledge of the ocean, both knowledge produced by scientists and that gained by recreationalists drawn by play into learning about the sea.

Recreation and the Cultural Discovery of the Ocean

Recreation has been, and remains, one of the key ways that people, both scientists and others, interact with the ocean. Beach visits, scuba diving, recreational fishing, boating, shell collecting, and other such activities allowed individuals to know the ocean on their own terms, just as so many people, following the example of Sierra Club founder John Muir, set out to encounter terrestrial places like Yosemite. Their knowledge of Yosemite was conditioned both by what they read and heard from others, including experts, and also by what they themselves saw and felt. In a parallel way, recreation proved a particularly important aspect of the late eighteenth-century cultural discovery of the seashore and the subsequent discovery of the ocean's depths.

Starting in the late eighteenth century, first Europeans and then Americans gained greater awareness of the sea. This consciousness derived from sources such as the results of scientific investigations, maritime writings, and the expansion of maritime industries, as well as recreational activities.[17] Increasing knowledge of the oceans emerged from both work and play—but, in many cases, new knowledge derived from activities that bridged those two categories. Just as work at, on, and in the sea changed nature, so too did play.

Cultural historian Alain Corbin described the discovery of the beach, when "the desire for the shore . . . swelled and spread."[18] His story involved tourism, geology, religion, medical practice, and fashion. The first step toward admiration of the sea started with the mid-eighteenth-century grand tour visits by French and English tourists to Holland. Tourists sought the sense of the sublime that arose from direct experience of the sea at the strand, but storms remained pictorial elements viewed from land rather than immediate dangers. Tourism, then, contributed "an apprenticeship in viewing the sea-shore."[19]

The coast also became a site for study by geologists and a place where others could reflect on the new scope of time proposed by scientists. Cliffs offered three-dimensional spectacles of time that suggested a disconnection between the histories of people and of the earth. The idea of a very ancient earth, indifferent to human presence, achieved sublimity.[20]

The seashore's appeal was not limited to reflection and aesthetics, but depended heavily on bodily sensations and the expansion of professional medicine. Sea bathing came into vogue in the context of visits to the seashore for health reasons. Sea air was healthful and bathers sought cures for melancholy, spleen, and anxiety through strictly regulated hydrotherapy. Under the guise of therapy, genteel bathers experienced a new range of bodily sensations. The fashionable society that grew up at seaside resorts such as Brighton eventually replaced the medical reasons for spending holidays at the beach with social motives.[21] In the 1840s, railroads extended to the coasts and unleashed crowds on the beaches in Britain.[22] American discovery of the shore, which followed closely, likewise depended on the confluence of railroads, therapeutic bathing, the moral benefits of natural history collecting, and desirable society.[23]

Corbin's story of the discovery of the seashore argues for a strikingly new cultural image of the ocean arising in the late eighteenth century. Fear and repulsion of the sea, inspired by biblical imagery and classical associations between the ocean and decay, gave way to admiration, then pleasure. By 1840, "the modern beach was born."[24] This new view of the seashore and near-shore waters was the product of a combination of work and play. Geology and medicine intersected with fashion and religion, serious approaches to the sea intersected with frivolous ones. The new recreational activities that brought wealthy people to the seashore were, to others, sources of employment and the foundation for a new coastal economy based on second homes, an extension of the social season, and, later, middle- and working-class tourism.

The new conception of the beach laid the foundation for new cultural understandings of the ocean itself. Dawning interest in the deep sea came from scientists, whalers, and submarine telegraph entrepreneurs—but also from novelists, readers, yachtsmen and -women, and amateur naturalists. In their hands and minds, the open ocean shifted from highway to destination.[25]

There is no doubt that work was a critical category for understanding the mid-nineteenth-century discovery of the ocean. Whaling and sealing brought mariners to the ends of the earth, often in advance of official national exploring parties, and decimated marine mammal populations everywhere they went.[26] Shipping grew in step with the massive expansion of industrial capitalism, carrying raw materials and manufactured products from and to the far corners of the globe. Confident and comprehensive knowledge of the sea's tides, currents, and contours made possible not only the quickening tempo of global commerce, but also new forms of transportation, such as the steamship, and new forms of communication, specifically the transatlantic telegraph. Understanding the ocean and its depths provided maritime nations with control that underlay the ability to dominate lands and cultures connected by major oceans. Knowledge of the oceans enabled imperialism.[27]

As much as the nineteenth-century discovery of the deep ocean owed to work, it owed at least as much to the popularity of seaside holidays, amateur natural history collecting, popular writing, and the brief, mid-century aquarium craze. Yachting transformed from a rare, aristocratic activity into a sport and pastime pursued by upper- and middle-class recreational sailors in Europe and the United States. Maritime novels were central; the first generation of maritime novelists at mid-century, including Richard Henry Dana and Herman Melville, numbered among the first respectable, middle-class people to seek the experience of voyaging. Novelists contributed to the new cultural construction of the deep sea. So did scientists, who wrote voyage narratives and popular works to introduce newcomers to marine natural history, tide-pool exploration, aquarium keeping, and the new scientific discoveries about the ocean's depths. Readers came to know the ocean through these works, often in conjunction with their own direct experiences of beach visits or boating.[28]

Recreation, in short, formed an inextricable element of both the eighteenth-century discovery of the beach and the subsequent, and connected, discovery of the deep sea and open ocean. These discoveries were cultural and involved recreation, yet were firmly tied to industrial, mercantile, and political uses of the ocean. It would be difficult, and probably pointless, to disentangle work from play for these episodes.

Science, Play, and Twentieth-Century Ocean Science

Recreation remained a central category for knowing the ocean long after the initial Western "discovery" of the beach. It remained so even for modern, professional sciences of the sea. As oceanography and marine biology developed, from mid-century forward, work and play continued to be intertwined. For example, scientists who worked for the U.S. Fish Commission in the last quarter of the century combined their summer fieldwork with family holidays, just as mid-century gentlemen-naturalists had done before. Spencer Fullerton Baird used summer collecting visits to the New England coast as opportunities to bring his wife and daughter to the seaside for healthy respites from the heat of the capital city. The establishment of the Marine Biological Laboratory at Woods Hole, Massachusetts, followed from the habit of many biologists, including but not limited to Fish Commission workers, to spend summers there, with their families, to gain ready access to marine organisms.[29]

Several decades later, the Woods Hole Oceanographic Institution, whose founders were attracted to the town's location close to deepwater, made a revealing choice in the design of its first research vessel. Since the establishment of the International Council for the Exploration of the Sea, oceanographic research was conducted from vessels with characteristics of working boats, mainly resembling trawlers for bottom-fishing. Member countries each committed to provide a vessel to participate in quarterly hydrographic cruises. Norway's vessel, the *Michael Sars*, for example, was designed and purpose-built for scientific research. Because of the type of work involved—deploying instruments over the side and working offshore in deepwater and in all weather conditions—its designers started with the hull characteristics and rig of a modern steam trawler.[30] Woods Hole director Henry Bryant Bigelow, despite his awareness of the Council's work and his own experience using modern, motorized vessels owned by the U.S. Fish Commission, chose to design a research vessel that was at least as much yacht as workboat. Bigelow justified his decision by projected savings of fuel costs from employing wind power and the stability offered by a sailing vessel. But, as Susan Schlee wrote in her history of the research vessel *Atlantis*, both Bigelow and Columbus O'Donnell Iselin, the man chosen to be the ship's master as the vessel was designed, "had an aristocratic preference for sailing . . . came from well-connected families and had sailed all their lives."[31] Their pursuit of oceanography was intertwined with their love of the sea.

The compatibility of seaside holidays and marine science extended into the post–World War II period, a time in oceanography's development that has been strongly associated with work. Historians have documented the meteoric growth of the ocean sciences and demonstrated its close connections to military needs and patronage.[32] While it is quite true that oceanography, especially physical oceanography, was allied to strategic uses of the ocean, the connection between recreation and science of the sea nevertheless remained strong through the 1950s and 1960s.

"Oceanography Is Fun" declared the title of a chapter of Elizabeth Shor's history of the Scripps Institution of Oceanography that covered expeditions in the 1950s and 1960s. Shor was married to the oceanographer George Shor and, like many Scripps wives, was active in campus social activities. The theme of oceanography as fun was the focus of an original musical revue created and performed in 1960 by the women's group at Scripps, featuring "mermaids and grunion hunters, sons of the beaches, girls from the 'Friendly Islands,'" and a dance titled, "Flip, Flop, Flip," performed in swim fins.[33] While at sea and stationed on Pacific islands, Scripps oceanographers integrated into their scientific work diving for pleasure, enjoying sunsets, and collecting art and objects from natives. As Roger Revelle put it, "At sea they gathered records and samples and specimens for science; ashore they gathered kava bowls, tapa cloths, cowry necklaces, vicuña rugs, carvings, 'antiquities'—and memories."[34] Diver-scientists not only installed wave-recording instruments off Enewetak in 1952 in preparation for the detonation of Mike, the first thermonuclear bomb, but they also explored coral reefs and local shipwrecks.[35] Willard Bascom, in his autobiography subtitled "Adventures in Oceanography," recalled returning from his first swim over a coral reef "with dozens of tiny cuts on my hands, a sunburned backside, and a firm resolve to return with a color movie camera."[36] This he did, filming reefs, fish, sharks, and men working underwater, "to record the scene for our friends at home."[37]

Indeed, "remote ports became collectors' items" and scientists referred to what were officially called "expeditions" as "sea trips" or "cruises."[38] A relatively early Scripps postwar expedition, deployed in 1951 to the Gulf of Alaska, was officially named the "Northern Holiday Expedition." In naval parlance, a holiday is an area in which work has been left undone, but oceanographer-planners relished the double meaning—then made a judicious decision to call the next expedition "Shellback" instead of "Southern Holiday," out of sensitivity for funding sources.[39] Still, oceanographers of the 1950s and 1960s were unrepentant that their love of oceanography was tied to the fun they had doing things near, on, and under the sea.

Undersea Exploration: Work or Play?

If recreation remained firmly tied to marine biology and oceanography as these fields developed, play emerged as a central and integral component of what scientists, engineers, and popular writers of the 1950s and 1960s called "undersea exploration." While the ocean's great depths became the theater in which the United States and Soviet navies played their version of blind man's bluff, the continental shelves were expected soon to yield such valuable commodities as food, mineral resources, and even living space. Especially for that part of the sea increasingly accessible for visits by human divers, the links between work, scientific work, and play in the ocean remained tight.

As with other aspects of ocean science, the military contributed significantly to the development of diving during and after the Second World War. Underwater sabotage and salvage were conducted using diving bells for hundreds of years and helmeted suits starting in the nineteenth century. Effective closed, helmeted suits came into regular commercial use in the late 1830s with Auguste Siebe's rig, which was improved and adapted by a number of other inventors, military officers and doctors, and commercial salvors. Work done using such rigs was indeed work: underwater construction, demolition, salvage, and reconnaissance. During the Second World War, naval frogmen made significant contributions to the war effort using closed circuit equipment that utilized chemicals to remove carbon dioxide from a diver's exhaled breath. The absence of bubbles to reveal the diver's presence provided a tactical advantage, but the system limited divers to only about thirty feet. Diving with modern gear remained in the hands of military and commercial specialists until French naval officer Jacques-Yves Cousteau and his partner, the engineer Emile Gagnan, invented their famous Aqua-Lung and opened the undersea world to recreationalists as well.[40]

Before that happened, most ocean enthusiasts expected exploration of the ocean to parallel that of outer space. They predicted that exploration would be government-funded, involve military personnel, hardware, and motives, and employ technology of the sort that was firmly out of reach of individuals. Security and profit were of paramount concern. At a time when experts predicted the speedy arrival of limits of agricultural production and consequent world hunger, the continental shelves began to be perceived as usable space for farming and massively underexploited in terms of the amounts of seafood and types of species harvested. Experts anticipated that oil and mineral extraction would soon be economically viable. The deep sea was seen as a potential dump site for atomic and other wastes, while the movement to find peaceful uses of atomic power suggested the idea of using nuclear energy to create artificial upwelling zones to bring nutrients to the surface in otherwise nonproductive parts of the sea.[41] Such new uses of the sea—mining, dumping, farming—were predicated on the development of a new, high-technology sector akin to the emerging aerospace industry. This set of expectations for the ocean clearly involved more work than play.

Then the trajectory for ocean exploration changed dramatically, and diverged sharply from that for space exploration, with the possibility for ordinary people to visit the undersea world. Before scuba, skin diving provided small numbers of people with recreational access to the ocean. In some cultures, of course, skin diving has been for millennia an integral part of subsistence or commercial activity, often both, such as by Chinese pearl divers, Greek sponge divers, Bahamian spear fishermen, or the Japanese women divers of Ama. In many of these places, divers were specialists rather than members of the general population. As a popular hobby, skin diving mainly took hold in warm Mediterranean waters, although in the United States some pioneer skin divers used spearfishing as a way to put food on the family's table during the Depression.[42] In the 1930s, skin divers explored the French Riviera, where the American expatriate Guy Gilpatric learned to dive and wrote his guide to the new sport, *The Compleat Goggler*.[43] Scuba diving likewise emerged from the beaches of France, where Cousteau and Gagnan invented the Aqua-Lung in 1943.[44]

Although scuba began in France, in the words of pioneer diver and diving historian Eric Hanauer, "the beaches of California were its incubator."[45] By 1949, when the San Diego Bottom Scratchers skin diving club was featured in an article in *National Geographic*, there were an estimated eight thousand skin divers in southern California.[46] By 1965 a popular book on skin and scuba diving claimed over six million skin divers in the United States, whose numbers included men, women, boys, and girls.[47] Initially, the main activity of skin divers was spearfishing, although as the sport grew in popularity, more divers began to hunt their quarry with cameras or simply enjoy seeing the underwater world for themselves. A close connection between recreational diving and science also dates from the early years of skin diving.

Diving and Science

The first generations of marine scientists, for the most part, did not turn to diving to gather specimens or information.[48] Diving became a scientific tool only in step with its popularization as a recreational activity. One early diving club well exemplifies both diving history and the foundational connections between diving and science. The San Diego Bottom Scratchers organized themselves in 1933, three years before Cousteau began goggling in the Mediterranean. The club's name derived from its initiation requirement: to capture three abalones on a single breath, to catch a ten-pound lobster, and to seize a horn shark with bare hands (all of which were significantly easier to do then compared to today). The horn was removed from the shark and hung from the zipper pocket of the diver's swim trunks, where it scratched sand when he swam along the bottom, hence the club's name.

Although most Bottom Scratchers spent their time underwater hunting, a few members forged a connection with Scripps Institution of Oceanography that led to enthusiastic efforts to apply diving to science. The relationship with Scripps began in 1945, after members who shot and captured a fish that weighed over one hundred pounds learned from Scripps ichthyologist Carl Hubbs that it was a broomtail grouper from a small population that might have been stranded in local waters by a warm episode. The Bottom Scratchers agreed to forgo hunting this population (although other divers did not). In subsequent years, two men who went on to become the first diving officers at Scripps joined the club, Jim Stewart in 1951 and Conrad Limbaugh (its first diving officer) in 1953. Hubbs himself joined in 1955.

Hubbs was unusual among scientists for having tried to incorporate diving in his research when only the helmeted apparatus was available. He had used a diving helmet in the 1940s and 1950s, both diving himself and employing divers to help. He strongly promoted the use of diving by marine scientists, especially after scuba technology became available. Eugenie Clark, who went on to become a well-known ichthyologist and wrote popular books about her experiences diving for research, made her first dive as a graduate student working with Hubbs.[49]

Another early example of connections between diving and science can be found in the career of the Austrian filmmaker and adventurer Hans Hass. He discovered skin diving while on holiday in Antibes, when he saw Guy Gilpatric spearfishing. He returned home to Vienna to pursue graduate study in zoology, made a diving helmet, and dove in the Danube River and the Adriatic Sea. In 1939 he published a book on underwater hunting, then a magazine serialized an account he wrote of a trip he made to the Caribbean to dive, hunt, and take underwater pictures. From that point, although he earned his doctorate in 1943, Hass's career was more focused on filming the underwater world than conducting research. He began diving with an oxygen rebreather, a technology that predated (and was superseded) by scuba, which allowed him to stay underwater long enough to film moving pictures. A film he made about the Red Sea won an international prize in the documentary category at the 1950 Venice

Film Festival. His films and television productions, as Cousteau's would later, raised popular interest in marine biology.[50] Hass's work illustrates the central role that photography played in the development of diving, scientific diving, popular interest in the undersea world, and the intersections between all of these.

The connection between diving and science strengthened dramatically with the advent of scuba. Cousteau first imported the Aqua-Lung into the United States in 1949, to a sports store in Westwood, California, near the University of California, Los Angeles, campus. One of the first units was purchased by a professor, Boyd Walker, at the urging of then graduate student Andreas Rechnitzer, who was pursing his Ph.D. at Scripps (at the time, Scripps was part of UCLA). Rechnitzer and fellow graduate student Conrad Limbaugh together figured out how to use the device and brought their enthusiasm for it to the Scripps campus when they arrived there in summer 1950. Scuba was employed by Scripps researchers first for a series of studies to determine the effects of kelp harvesting on fish populations and the causes of declines of kelp beds.[51]

Another early scientific application of scuba essentially created the new field of underwater archaeology. Rechnitzer, who became much more famous for his scientific leadership of the only successful effort to put people at the bottom of the Marianas Trench in 1960, recovered ancient Indian bowls and artifacts from La Jolla Canyon in the 1950s. Divers had begun finding artifacts along the southern California coast early in the decade. A shallow stretch about a mile south of Scripps proved a particularly rich source, from which hundreds of small mortars were collected. Although Scripps never became a center for underwater archaeology, some of its diving scientists dabbled in the new field as they experimented with scuba for their own research. Hubbs, who was quite catholic in the breadth of his research interests, got involved in a long archaeological study of middens left by early southern California inhabitants.[52] The new field of underwater archaeology, established on a scientific basis by its acknowledged founder, George Bass, owed its existence to the new access to the undersea world offered by scuba and, later, small, submersible research vehicles.[53]

From the perspective of the early 1950s, many Scripps scientists were enthusiastic about the possibilities offered by scuba technology. Members of the Bottom Scratchers were sometimes recruited to help with underwater work that was part of Scripps research projects. Indeed, in a few instances, the divers themselves became the research focus, when scientists studied the divers' physiology while exercising and holding their breath underwater.[54] Rechnitzer recalled being told by Wesley Coe, a marine invertebrate zoologist, "Son, you go on out there. You will learn more in an hour [underwater] than you will learn in ten hours with books."[55] Examples of research done at Scripps employing scuba include Wheeler North's long-term study of the effects on the marine environment of the Tampico oil spill in the late 1950s, which affected an area sixty miles south of Ensenada.[56] Ecologist Edward W. (Bill) Fager conducted studies of the community structure of an underwater area dubbed "Fager's half acre." He observed naturally occurring, common invertebrates, and he also installed artificial "rocks" in underwater sandy areas to monitor community development in new regions.[57] Geologist Francis P. Shepard began his investigation of the submarine canyon off Scripps using sounding lines and helmeted divers, but eagerly tried new technologies as they became available, including scuba, Cousteau's diving saucer, echo sounders, and underwater cameras.[58] Douglas L. Inman employed both scuba and instruments with the Shore Processes Study Group to investigate the mechanics of beaches.[59] Application of scuba was not limited to the waters off Scripps. The 1950 MidPac Expedition was cosponsored by the Sea-Floor Studies group at Navy Electronics Laboratory, a group of scientists described by Scripps historian Elizabeth Shor as "a lively one, all Scuba divers who often sought their geology with the aid of swim-fins."[60] Between 1965 and 1978, according to an estimate by Jim Stewart, at least thirty dissertations relied on diving as a central research tool.[61]

Scripps became ground zero for scuba, its program serving as a model for scientific diving in other U.S. universities. The system of diving protocols and instruction developed there evolved into sport diving courses and, thus, formed the basis for recreational as well as scientific diving training in the United States. Limbaugh and Rechnitzer devised such techniques as the buddy system, buddy breathing, and ditch and recovery. These techniques became part of a written curriculum taught at the University of California in 1952, which was the first civilian diving course offered in the United States. Before that time, informal demonstrations of scuba equipment took place among friends, colleagues, and dive shop operators. The underwater deaths of students at University of California campuses other than Scripps motivated the university system to formalize training and certification of university-affiliated divers. Limbaugh, who became the first diving officer at Scripps, discussed diving training with E. R. Cross, founder of the Sparling School (one of the first commercial diving schools) and author of a 1951 manual on "Underwater Safety." Cross was invited to serve on a panel for developing the first public instructor training and certification program, and the three people who created the Los Angeles County program—Al Tillman, Ramsey Parks, and Bev Morgan—were sent to take one of the first formal diving courses offered by Limbaugh at Scripps. Before commercial classes were available through dive shops, Wheeler North started a public diving program through San Diego's Parks and Recreation Department. At Scripps, sport diving literally added a new dimension to science by giving researchers access to the undersea environment, while diving scientists, in turn, contributed substantively to the emerging practice of recreational diving.[62]

Other developments further confirm the profound influence exerted through Scripps on recreational and scientific diving. Scripps divers and scientists had a hand in testing, and suggesting materials for, the first wet suit, invented by Hugh Bradner.[63] Jim Stewart, who became the Institution's diving officer after Limbaugh's tragic death cave diving in France, continued to develop techniques for diving, and especially for scientific diving, that have remained in use. An avid free diver and spearfisherman, Stewart met Limbaugh at a spearfishing meet and Limbaugh convinced him to volunteer with kelp studies and the diving training course. Stewart made his career at Scripps, assisting in faculty research and collecting fish for display at the institution's public aquarium. He was responsible for expanding Scripps's diving curriculum across the University of California system, developing guidelines for diving programs at all campuses, and writing the first university diving safety manual, which was published in 1960. Said Stewart of the safety standards and procedures he wrote for scientific diving, "The people that started scuba were water people. We trained the rest."[64] The connections found at Scripps between recreation, diving, and science were replicated in other institutions and in the careers of many individuals.

An Accessible Arena

Recreational diving rendered the undersea world accessible to hunters, scientists, shutterbugs, nature watchers, and technology enthusiasts. Although underwater exploration in the late 1940s still had strong military associations, diving did not long remain the work-related pursuit it was in the days of helmeted commercial diving and military underwater swimming. By the mid-1950s, with the introduction of scuba, training, and wet suits, diving was en route to becoming an accessible pastime for many people, male and female, young and old. In 1956 there were 132 dive clubs in California, 32 in New York, and 24 in Florida.[65] By 1965 more than 650 dive clubs had been formed in almost every part of the United States.[66] An Aqua-Lung cost $160 in 1954, when a new, twelve-inch color television cost $1,100; *Popular Science* magazine published an article explaining how to build a "diving lung" from inexpensive parts, including oxygen tanks and a demand valve.[67]

As diving opened up recreationally, many women took up the sport. E. R. Cross recalled that there were two women in his first class at the Sparling

School, one of whom became a renowned shell collector.[68] Free diver and commercial fisherman Dottie Frazier was the first woman to take the scuba instructor's course. She resented the militaristic feel of the program but got a job teaching scuba and made a career out of diving by buying a dive shop and making custom wet suits.[69] Diving reminiscences are full of tales of diving pioneers making their own wet suits, but women divers did not have to seek out custom suits for long. Women's models were advertised within the first few years that wet suits were available commercially.[70]

Frazier was a charter member of a women's dive club, the Beach Neptunes. While the Bottom Scratchers remained an all-male club, wives and girlfriends of its members formed their own diving group, the first women's club, called the Sea Nymphs. Like many of the Sea Nymphs, Zale Parry, the third woman to be certified to teach scuba in the United States, got her start underwater by waiting for her date and his buddies to get cold while diving, then using air left in their tanks to dive. "It was a long time before I dived a fresh bottle. I probably made the most free ascents in the world; they were just part of the sport to me."[71] Her subsequent career, as the leading Hollywood underwater stunt girl, inspired many men and women to take up diving in the 1950s and 1960s. Her career also highlights the impact of film for gaining and communicating knowledge of the ocean in the 1950s and 1960s. Parry was best known for her work in the popular *Sea Hunt* television series, in which she sometimes starred "topside" but underwater stood in for all the woman actors. She also did stunt work for movies, such as *Boy on a Dolphin*, in which she doubled for Sophia Loren. In 1954 Parry became the first woman to dive below two hundred feet, an achievement that put her on the cover of *Sports Illustrated*.[72] The feat was sponsored by diving companies intent on demonstrating to the public that anyone could dive.

Women's participation in diving helped transform the undersea realm into an accessible arena. Besides Parry on television, two women scientists contributed notably to the idea that ocean exploration could be undertaken and enjoyed by ordinary people. Both Rachel Carson and Eugenie Clark were popular writers as well as scientists, and, as historian Gary Kroll demonstrates, their work drew the attention of their readers to the unfamiliar, mysterious, and exciting undersea world.[73] During the years that Carson was writing *The Sea Around Us*, she took advantage of an opportunity to dive with F. C. Walton Smith, director of the new University of Miami marine station. Carson's work synthesized and brought to the public for the first time the scientific knowledge gained about the ocean as a result of the Second World War.[74]

Carson's literary agent, Marie Rodell, also represented Eugenie Clark, whose books *Lady with a Spear* and *The Lady and the Sharks* provided readers with adventurous, popular accounts of her pioneering underwater research on fish, invertebrates, and sharks. Clark, an avid swimmer from childhood, began her career as Hubbs's graduate assistant. With his encouragement she took up free diving and later spearfishing as a means for observing and collecting specimens for her research. She went on to conduct extensive fieldwork using skin diving equipment in the Caribbean, Hawaii, the Red Sea, and Micronesia. Later she adopted scuba for her work, especially after she established the Cape Haze Laboratory in western Florida in 1955 (now Mote Laboratory.[75]

That Clark's motive for skin diving was as much sport as work is apparent from her books, starting with her description of the moment she resolved to learn to spearfish. She watched Vernon Brock, director of fisheries in Hawaii, shoot a huge, green parrotfish underwater. Her view of the spectacle was interrupted as she was forced to take several breaths during the time Brock remained submerged, hunting the fish: "I vowed that someday I would learn to use a spear and catch fish this way in their own environment. In comparison, how dull it is to sit up on the airy world and pull a fish out of the water on a line!"[76] To Clark, science itself was as much fun as work, especially the collecting. En route to a job in Guam, during an unplanned, two-day stopover at the Pacific island of Swajalein, Clark orchestrated a test of the application of rotenone poison on a tide pool as a technique for collecting fish. To the naval personnel and their families stationed on the island, "the idea of a scientific fishing venture had the appeal of a picnic party. . . . We had a grand time catching gobies and blennies, those agile leapers that jumped out of our pool to escape its fouled waters and went skipping over the dry reef to other, healthier tide pools, with several Naval officers hot on their trail."[77] Not only was this kind of collecting enjoyable for Clark, but it also entertained nonscientists on the island and, later, readers of her popular books.

Clark's celebrity status and popularization efforts contributed to the democratization of ocean exploration by presenting diving as an activity for ordinary people, not only for scientists.[78] She assured her readers, "Once you are familiar with the sea's dangers and know where to expect them and how to avoid them, you can roam with safety and assurance in a world of wonder that otherwise you will never really know."[79]

Diving very often became a family activity. Clark's new husband, on their honeymoon, bought a mask and spear gun and joined in her underwater activities. Children learned to dive from their parents.[80] One popular diving book enthusiastically reported that divers "rang[ed] from boys and girls in their teens to men and women in their fifties and older."[81] Nor was diving seen as particularly dangerous for pregnant women. Cross, in his regular

"Technifacts" column in *Skin Diver* magazine, devoted a column to diving during pregnancy, "just to start the dialog."[82] Clark, whose husband was a medical doctor, dove during all her pregnancies, believing the relative weightlessness of the underwater environment to be beneficial to both her and her baby.[83] While five months pregnant, the marine botanist and underwater explorer Sylvia Earle dove at nearly one hundred feet, having been transported to the site by a submersible with lock-out capability. She was able to stay an hour and a half, longer than she had ever remained at that depth, because the submersible had its own decompression chamber. In recounting the experience, Earle recalled giving her "undivided attention" to the plants she went there to study, but on the way back up, in the decompression chamber, she reflected on what her child would think, years later, when old enough to be told about the experience. She commented offhandedly that "doctors I had consulted foresaw no difficulties, and there were none."[84]

The careers of Rachel Carson, Eugenie Clark, and Sylvia Earle emphasize the extent to which the general public learned about the ocean through popular writings by scientists in the post–World War II world. As Hans Hass's filmmaking and Zale Parry's stuntwoman achievements suggest, books were joined, and possibly superseded, by photography and film as means for helping non-divers learn about the mysterious undersea world. The underwater view perhaps offered viewers vicarious experience and very likely inspired some to take up diving to encounter the ocean themselves.

Before scuba or even the rebreathers that preceded scuba technology, efforts were made to photograph and film underwater. Both recreational and scientific divers were instrumental in the development of underwater photography, and the results paired science and spectacle in a way that scuba would later go on to do. In the 1920s, scientist-curator Roy Waldo Minor visited Andros Island, in the Bahamas, and there discovered a device called a photosphere, invented by John Ernest Williamson to provide photographers with access to underwater environments without getting wet. Camera operators using the long tube with a viewing chamber mounted on the end filmed underwater scenes for one of the first film versions of Jules Verne's *Twenty Thousand Leagues Under the Sea* in 1916. Minor used the viewing chamber for making detailed observations of a coral barrier reef in preparation for a major exhibit installed in the American Museum of Natural History that opened in 1930 and another on South Pacific pearl divers that opened in 1941.[85] After the war, photographers started bringing cameras into the water.

As with the adoption of scuba for scientific research, diving clubs became important sites for technical development for underwater photography. Lamar Bowen, a member of the Bottom Scratchers from 1943, applied his expertise from his own photography studio when he went underwater. He started shooting still photos, then moved on to movies, benefiting from help fellow members gave him in constructing waterproof camera housings. His subsequent thirty-year Hollywood career as a leading underwater filmmaker got started when he was asked to shoot the underwater scenes for the 1955 Jane Russell film *Underwater!* In addition to *Sea Hunt*, which he helped to originate, Bowen's other credits include the *Flipper* television series, *Old Man and the Sea*, *Day of the Dolphin*, *Around the World Under the Sea*, and several James Bond films.

Scientist-divers, too, took cameras underwater. Hans Hass is a good example, as is the consummate ocean explorer–showman, Jacques Cousteau. Sylvia Earle likewise joined forces with National Geographic on several occasions to film her ocean exploration. Photography, especially film, acquainted audiences, especially people who did not dive, with the underwater world. Films, and the exploits of the stars such as Zale Parry, also encouraged people to try ocean exploration for themselves. Just as for writer-popularizers such as Eugenie Clark or Rachel Carson, divers and diving scientists who chose film as a medium for their work helped reshape the image of the sea from a forbidding and dangerous place into a place accessible to many people, safe to explore, and amenable to understanding by direct experience made possible through scuba technology.

Explorers and Scientists

The intersection between scientific and sport diving involved more than training and equipment, and also more than popular acquaintance with the ocean through movies and books. For those who chose to dive, it involved a personal, physical encounter with the sea. The meaning of that encounter varied with the diver, but most understood their time spent underwater as a form of exploration of undiscovered territory. Herb Pfister, author of a *Popular Science* magazine article about how to build a diving rig at home, touted the new sport: "For a brand-new sensation, a feeling of really being out of this world, try the latest thing in water sport."[86] A 1965 popular book on skin and scuba diving asserted, "Those with a love of adventure and exploration have been quick to take advantage of the thrills offered by this new key to the unknown."[87] Through the trope of exploration, divers could know the ocean for themselves.

Another popular manual contrasted the views of the diver and the nondiver: "The non-diver stands landlocked at the edge of the sea . . . He cannot conceive the excitement, the peace, or the beauty of that which lies hidden from his eyes.

Magnificent canyons and valleys, vast deserts of pale blue sand, enchanted forests of dwarfed trees and exploding colors and shapes reveal themselves only to the diver's eyes."[88] This privileged view of the ocean took a diver "out of his natural environment."[89] The physical experience was an inextricable part of the feeling of achievement for the diver. "You'll find in the underwater world . . . the joy of a new and unique sensation . . . and a feeling of pride at your own part in conquering a world which has baffled man for centuries."[90]

Professional scientists were equally affected by their experiences underwater. As oceanographers Willard Bascom and Roger Revelle explained in 1953, a scientist undersea using scuba "becomes part of the medium. Subject to the same forces, he partakes of the feelings of the underwater animals as he is swaying with passing waves and drifts along, weightless, with the ocean currents."[91] As for the recreationalist, the scientist believed that the bodily feeling of being underwater was integral to the value of the experience.

For both scientists and recreationalists, diving, as a way of knowing the ocean, combined the bodily experience of the undersea environment with a scientific perspective. Indeed, it seemed impossible to dive and not practice science. "Everyone who goes underwater becomes an amateur scientist," declared science fiction author Arthur C. Clarke, who in 1960 narrated a visit to an undersea resort that he imagined in business within the decade. Guests at the underwater hotel on a coral reef would don scuba gear and, with a wave to those watching from underwater observation windows, follow their guide on an excursion to see and interact with spectacular marine creatures, traveling far from the hotel using novel propulsion devices.[92]

Clarke, who was more famous for his imaginative books about space travel, became caught up with the 1950s enthusiasm about the ocean. He learned to scuba dive, joined the Underwater Explorers Club, and became so enamored of the sport that he moved to Ceylon (now Sri Lanka) in 1956 to pursue this interest. He opened a dive school that was still in business when the 2004 tsunami hit.[93] Before he published his blockbuster *2001: A Space Odyssey*, some of Clarke's writing focused on the sea. He wrote a nonfiction work, *The Challenge of the Sea*, which predicts future technologies that would enable new uses for the sea, as well as another, *Voice Across the Sea*, that celebrated the achievements of laying telegraph and telephone wires across oceans.[94] A nonfiction children's book, *Boy Beneath the Sea*, was the product a partnership with a photographer to excite and inform young readers about the prospects of diving to spearfish and hunt for treasures from wrecks.[95] Fictions works from this period include *The Deep Range*, a novel about a future submariner who farms the seas, and *Dolphin Island: A Story of People of the Sea*, about a boy whose life was saved by dolphins, who take him to an island where a professor is trying to communicate with these marine mammals.[96]

Clarke's interest in diving seems to have sprung from his desire to feel weightlessness akin to that experienced by space travelers. His fascination for the undersea world underscores parallels between space and the ocean in terms of cultural perceptions and attitudes in the postwar decades. Other space enthusiasts embraced the ocean as well, including the famous rocket scientist Wernher von Braun, whom Clarke introduced to the sport of skin diving. In his introduction to Clarke's *The Challenge of the Sea*, von Braun naturalized both space and ocean exploration by declaring, "We on earth can consider the bottom of the sea as man's point of departure on his extremely long trip to outer space. Life began in the depths of the sea, and through eons moved upward toward land. Today, after a brief pause, he has means for continuing his journey from the land upward."[97] Others voiced this evolutionary argument for the inevitability of human exploration as well. The philosophical marine biologist Sir Alister C. Hardy, in a 1960 lecture to a British diving club at Brighton, posited a theory that humans went through a long aquatic phase during evolution.[98]

Clarke and von Braun were fascinated with science and technology as much as with the frontiers of space and the ocean. But they were not the only observers to assign recreational divers the role of amateur scientist. Diving instruction manuals routinely framed the sport as one of discovery, offering the chance to "set eyes upon a huge new canyon, discover a hitherto unknown animal." Such instructions took care to distinguish "adventure" from "exploration," such as when one instructor laid down "a challenge for every undersea adventurer to become an explorer. Here the geologist, the biologist, the geographer, and the sight-seer will find their greatest discoveries . . . The *real* age of discovery has just begun."[99] In a similar vein, the collaborators who produced a unique cookbook aimed at newly minted divers, *Bottoms Up Cookery* (see figure 7.1), argued that sport divers just past their checkout dive needed instruction in what prey to pursue, how to find and capture it, and, finally, how to prepare it. "The beginning diver is taught little about the sea life environment," lamented the authors, who advised hunters to "act like the sea creatures." They expected divers to learn what depths, water temperature, and types of areas fish and other prey inhabit, how they camouflage themselves, and what their feeding habits and migration patterns are. Divers should then "use the above knowledge and try to consider your underwater environment from the sea creature's point of view."[100] The diver in the cover illustration studies something on the seafloor with a magnifying glass. Even divers in search of dinner became scientists.

Figure 7.1

Cookbook authors present diving as a combination of recreation and pursuit of science, even when beginners do it. In this cover illustration, the diver studies the ocean to know how to find something good for dinner.

From Robert B. Leamer, Wilfred H. Shaw, and Charles F. Ulrich, Bottoms Up Cookery *(Gardena, Calif.: Mastergraphix, 1970).*

Conclusion

Writing the ocean's history requires paying attention to how people came to know the ocean. As environmental historians have learned from Richard White, work proves an essential category for knowing nature. Robert Kohler's scholarship demonstrates that, at times, play became work. In the mid-nineteenth century, work and play together contributed to the transformation of the ocean from highway to destination. In the 1950s and 1960s, the work-play of divers fostered new conceptions of the ocean as accessible and controllable, which intersected with the increasingly confident belief that new ways to exploit the sea were close at hand.

Historians of the field sciences tackling the question of how science has known parts of the global environment rely on categories of place, practice, and knowledge (particularly its production and circulation).[101] The opening of this essay raised the question of the uniqueness of the ocean. Ultimately, asking whether the ocean is a distinctive place is probably less productive than exploring how characteristics of the ocean may have shaped our understanding and use of it. What kind of place, then, has the ocean been?

The sea has been used and understood as an arena for war or diplomacy, a font of resources, a blank slate for discovery, a platform for movement, and a focus for reflection. At all times since the eighteenth-century discovery of the seashore, the relationship between people and the ocean has involved play as much as work. The examination in this essay of the postwar decades reveals that the category of play seemed intrinsic to the connections between people and the oceans that emerged at that time. To be sure, play did not eliminate the commercial cod fisheries in the 1990s, but commercial fisheries represent only one of many dimensions of human interaction with the sea. Many of these dimensions, not just fishing, have had an indelible impact on the ocean. This statement from the 1964 book *The Bountiful Sea*, written by Seabrook Hull, founder of the new field of ocean engineering, enumerates the ways he believed people and the oceans were intertwined: "Of the two great frontiers, space and the ocean, being opened up in the 20th Century, only the ocean is close, tangible, and of direct personal significance to every man, woman, and child on the face of the globe. Another war might be won or lost within its depths, rather than in outer space. It is a cornucopia of raw materials for man's industries, food for his stomach, health for his body, challenges to his mind, and inspiration to his soul."[102] Hull envisioned the sea as a source of underutilized resources and also as the site for a novel set of interactions between people and nature. The kind of place he described would, by definition, be changed by human actions, whether those were play or work.

Historians of the field sciences have demonstrated the value of examining the practice of science. This study of the intersection of play and the ocean suggests that scuba technology catalyzed the transformation of the forbidding arena of the ocean's depths into an accessible place, reachable by ordinary people as well as experts. Indeed, this technology was purported to give even recreational divers a scientific perspective during their time underwater. Diving enabled scientists to continue combining recreation with serious investigation of the ocean, while the new sport of recreational diving boosted the already-booming fields that composed the postwar marine sciences.

Recreational diving, undersea photography, television series such as *Sea Hunt* or movies such as *Flipper*—all these delivered knowledge of the oceans, no less than did work in, on, and under the ocean, including oil well drilling, scientific sampling, undersea construction, and emplacement and use of habitats for living and working underwater.[103] Working scientist-divers, however, persisted in understanding their activities as simultaneously work and play, not distinguishing between these categories. Knowledge of the ocean gained by divers, both scientists and recreationalists, leaned heavily on bodily experience, personal and direct, of the underwater world. The quality of knowledge produced about the ocean depended in part on experience and prior knowledge possessed by the diver. Although knowledge produced by scientist-divers appeared in science journals, recreational divers explored, learned about the sea, and contributed as much as scientists to the transformation of the underwater realm into an accessible place for exploration and discovery.

Notes

I wish to thank Gary Kroll for his repeated reminders not to forget play as a category in my thinking about the ocean. I also thank members of the HIST 302 seminar at University of Connecticut in spring 2007 for useful discussions of the ocean and of categories of work and play. This essay also benefited from comments and discussion by participants at the Knowing Global Environments conference held in Philadelphia in May 2007. Finally, I gratefully acknowledge a profound and happy debt to Robert Kohler for his approach to history of science and environmental history, which has guided my work for so long.

1. Helen M. Rozwadowski, *Fathoming the Ocean: Exploration and Discovery of the Deep Sea* (Cambridge: Belknap Press of Harvard University Press, 2004).
2. Richard White, "'Are You an Environmentalist or Do You Work for a Living?' Work and Nature," in *Uncommon Ground: Rethinking the Human Place in Nature*, ed. William Cronon (New York: Norton, 1996).
3. Richard White, *The Organic Machine* (New York: Hill and Wang, 1995); and Helen M. Rozwadowski, "Oceans: Fusing the History of Science and Technology with Environmental History," in *A Companion to American Environmental History*, ed. Douglas Cazaux Sackman (Malden, Mass.: Wiley-Blackwell, 2010).
4. White, "Are You An Environmentalist?" 172.
5. Ibid.
6. White, *Organic Machine*, x.
7. William Cronon, *Changes in the Land: Indians, Colonists, and the Ecology of New England* (New York: Hill and Wang, 1983); Cronon, *Nature's Metropolis: Chicago and the Great West* (New York: Norton, 1991); and White, *Organic Machine*.
8. White, "Are You an Environmentalist?" 174.
9. Alain Corbin, *The Lure of the Sea: The Discovery of the Seaside in the Western World, 1750–1840*, trans. Jocelyn Phelps (Berkeley and Los Angeles: University of California Press, 1994); and Rozwadowski, *Fathoming the Ocean*, 115, 126–127, 156.
10. Robert E. Kohler, *All Creatures: Naturalists, Collectors, and Biodiversity, 1850–1950* (Princeton: Princeton University Press, 2006), 67.
11. Ibid., 69.

12. Ibid., 47–90.
13. White, "Are You an Environmentalist?" 173.
14. Kohler, *All Creatures*, 90.
15. Helen M. Rozwadowski and David K. van Keuren, eds., *The Machine in Neptune's Garden: Historical Perspectives on Technology and the Marine Environment* (Sagamore Beach, Mass.: Science History Publications, 2004).
16. Rozwadowski, "Oceans."
17. Rozwadowski, *Fathoming the Ocean.*
18. Corbin, *Lure of the Sea*, 250–281.
19. Ibid., 19–56, on p. 35.
20. Ibid., 197–220. See also Martin J. S. Rudwick, *Scenes from Deep Time: Early Pictorial Representations of the Prehistoric World* (Chicago: University of Chicago Press, 1992).
21. Corbin, *Lure of the Sea*, 57–96.
22. Ibid., 250–281; Jean-Didier Urbain, *At the Beach*, trans. Catherine Porter (Minneapolis: University of Minnesota Press, 2003); and David E. Allen, *The Naturalist in Britain: A Social History* (London: A. Lane, 1976).
23. John Stilgoe, *Alongshore* (New Haven: Yale University Press, 1994), 295–367; and Lena Lencek and Gideon Bosker, *The Beach: The History of Paradise on Earth* (New York: Viking, 1998).
24. Corbin, *Lure of the Sea*, 281.
25. Rozwadowski, *Fathoming the Ocean.*
26. Michael F. Robinson, *The Coldest Crucible: Arctic Exploration and American Culture* (Chicago: University of Chicago Press, 2006); and Rozwadowski, *Fathoming the Ocean*, 37–66.
27. Philip Steinberg, *The Social Construction of the Ocean* (Cambridge: Cambridge University Press, 2001); and Michael S. Reidy, *Tides of History: Ocean Science and Her Majesty's Navy* (Chicago: University of Chicago Press, 2008).
28. Rozwadowski, *Fathoming the Ocean*, 1–37, 97–174.
29. Dean C. Allard Jr., *Spencer Fullerton Baird and the U.S. Fish Commission* (1967; repr., New York: Arno Press, 1978), 60–61, 164, 177–179; Frank R. Lillie, *The Woods Hole Marine Biological Laboratory* (Chicago: University of Chicago Press, 1944), 24–26; and Jane Maienschein, *100 Years Exploring Life, 1888–1988: The Marine Biological Laboratory at Woods Hole* (Boston: Jones and Bartlett, 1989).
30. Helen M. Rozwadowski, *The Sea Knows No Boundaries: A Century of Marine Science Under ICES* (Seattle: International Council for the Exploration of the Sea, 2002), 44–45.
31. Jeffrey P. Brosco, "Henry Bryant Bigelow, the U.S. Bureau of Fisheries, and Intensive Area Study," *Social Studies of Science* 19 (1989): 239–264; Susan Schlee, *On Almost Any Wind: The Saga of the Oceanographic Research Vessel Atlantis* (Ithaca: Cornell University Press, 1978), 16–17; and Schlee, *The Edge of an Unfamiliar World: A History of Oceanography* (New York: Dutton, 1973), 275.
32. Jacob Darwin Hamblin, *Oceanographers and the Cold War: Disciples of Marine Science* (Seattle: University of Washington Press, 2005); Naomi Oreskes, "A Context of Motivation: U.S. Navy Oceanographic Research and the Discovery of Seafloor Hydrothermal Vents," *Social Studies of Science* 33 (2003): 697–742; Oreskes and Ronald Rainger, "Science and Security Before the Atomic Bomb: The

Loyalty Case of Harald U. Sverdrup," *Studies in the History and Philosophy of Modern Physics* 31 (2000): 309–369; Oreskes, *Science on a Mission: American Oceanography in the Cold War and Beyond* (Chicago: University of Chicago Press, forthcoming); Rainger, "Science at the Crossroads: The Navy, Bikini Atoll, and American Oceanography in the 1940s," *Historical Studies in the Physical and Biological Sciences* 30 (2000): 349–371; Rainger, "Constructing a Landscape for Postwar Science: Roger Revelle, the Scripps Institution, and the University of California, San Diego," *Minerva* 39 (2001): 327–352; and Gary Weir, *An Ocean in Common: American Naval Officers, Scientists, and the Ocean Environment* (College Station: Texas A&M University Press, 2001).

33. Elizabeth Noble Shor, *Scripps Institution of Oceanography: Probing the Oceans, 1936 to 1976* (San Diego: Tofua Press, 1978), 373, 438, 465–466.
34. Shor, *Scripps Institution of Oceanography*, 374.
35. Willard Bascom, *The Crest of the Wave: Adventures in Oceanography* (New York: Anchor Books, 1988), 105, 136.
36. Ibid., 93.
37. Ibid., 104–105.
38. Shor, *Scripps Institution of Oceanography*, 375–378.
39. Ibid., 394, 399.
40. Robert F. Marx, *The History of Underwater Exploration* (1978; repr., New York: Dover Publications, 1990), 30–77; James Dugan, *Man Under the Sea* (1956; repr., New York: Collier Books, 1965), 19–21; and Sylvia A. Earle and Al Giddings, *Exploring the Deep Frontier: The Adventure of Man in the Sea* (Washington, D.C.: National Geographic Society, 1980), 30–31, 45.
41. Jacob Darwin Hamblin, *Poison in the Well: Radioactive Wastes in the Oceans at the Dawn of the Nuclear Age* (New Brunswick: Rutgers University Press, 2008).
42. Coles Phinizy, "The Old Men of the Sea," *Sports Illustrated*, August 23, 1965.
43. Guy Gilpatric, *The Compleat Goggler* (New York: Dodd, Mead, 1938).
44. For an account of the development of scuba, see Marx, *History of Underwater Exploration*, 86–96.
45. Eric Hanauer, *Diving Pioneers: An Oral History of Diving in America* (San Diego: Watersport Publishing, 1994), 11.
46. David Hellyer, "Goggle Fishing in Californian Waters," *National Geographic* 95 (1949): 615–632.
47. Gustav Dalia Valle, Benjamin S. Holderness, Charles M. Smithline, Arthur Stanfield, and Harry Vetter, *Skin and Scuba Diving*, Athletic Institute Series (New York: Sterling, 1965), 8.
48. Nineteenth-century scientists occasionally dove; one example was the French zoologist Henri Milne-Edwards, who dove off Messina, Italy, in 1844. Diving was not, however, a common practice among early marine zoologists of that era. Earle and Giddings, *Exploring the Deep Frontier*, 25; and Rozwadowski, *Fathoming the Ocean*, 33.
49. Eugenie Clark, *Lady with a Spear* (New York: Harper, 1953), 212–215.
50. Trevor Norton, *Stars Beneath the Sea: The Pioneers of Diving* (New York: Carroll and Graf, 1999), 198–216; and Dugan, *Man Under the Sea*, 198–199, 305.
51. Shor, *Scripps Institution*, 123–124; and Hanauer, *Diving Pioneers*, 96–99.
52. Shor, *Scripps Institution*, 134, 208, 335–336.

53. George F. Bass, *Archaeology Under Water* (New York: Praeger, 1966).
54. Hanauer, *Diving Pioneers*, 46–55; and Dugan, *Man Under the Sea*, 203–215.
55. Hanauer, *Diving Pioneers*, 100.
56. Ibid., 86–95.
57. Shor, *Scripps Institution*, 216–217.
58. Ibid., 273–276.
59. Ibid., 279–280.
60. Ibid., 286–287. See Dugan, *Man Under the Sea*, 203–212, for another review of Scripps scientists who used diving in research.
61. Shor, *Scripps Institution*, 134.
62. Hanauer, *Diving Pioneers*, 62–72, 90, 100, 108–117. An example of the resulting curriculum written by one of the founders of the Los Angeles County program is Albert A. Tillman, *Skin and Scuba Diving in Underwater Education: A Training Text for Institutions of Higher Learning* (Dubuque, Iowa: William C. Brown, 1962); p. 7 mentions Scripps's influence.
63. Carolyn Rainey, "Wet Suit Pursuit: Hugh Bradner's Development of the First Wet Suit," MS, Scripps Institution of Oceanography Archives, Reference Number 98-16.
64. Hanauer, *Diving Pioneers*, 118–127; and Shor, *Scripps Institution*, 127–136. Stewart also founded the American Academy of Underwater Sciences, originally intended to fend off federal government regulation of scientific diving by the Occupational Safety and Health Administration.
65. Dimitri Rebikoff, *Free Diving*, trans. Mervyn Savill, ed. Albert Vander-Kogel (New York: Dutton, 1956); and Hanauer, *Diving Pioneers*, 11.
66. Valle and others, *Skin and Scuba Diving*, 9.
67. Herb Pfister, "How to Build and Use a Diving Lung," *Popular Science*, July 1953, 160–169; and Hanauer, *Diving Pioneers*, 12–13.
68. Hanauer, *Diving Pioneers*, 62–72.
69. Ibid., 138–147.
70. Bruce Greenlaugh, former Navy SEAL diver, interview with the author, January 27, 2007; and Hanauer, *Diving Pioneers*, 10, 15.
71. Hanauer, *Diving Pioneers*, 148–159.
72. *Sports Illustrated*, May 1955.
73. Gary Kroll, *America's Ocean Wilderness: Cultural History and Twentieth-Century Exploration* (Lawrence: University Press of Kansas, 2008), 95–151.
74. Linda Lear, *Rachel Carson: Witness for Nature* (New York: Henry Holt, 1997); and Rachel Carson, *The Sea Around Us* (New York: Oxford University Press, 1951). An exceptional example of scholarship that focuses on women in postwar oceanography is Naomi Oreskes, "Laissez-tomber? Women's Work and Military Patronage in Twentieth-Century Oceanography," *Historical Studies in the Physical and Biological Sciences* 30 (2000): 373–392. See also Michael S. Reidy, Gary Kroll, and Erik M. Conway, *Exploration and Science: Social Impact and Interaction* (Santa Barbara, Calif.: ABC-CLIO, 2007), 194.
75. Clark, *Lady with a Spear*; Clark, *The Lady and the Sharks* (New York: Harper and Row, 1969); and Kroll, *America's Ocean Wilderness*, 124–151.
76. Clark, *Lady with a Spear*, 40.
77. Ibid., 71.
78. Reidy, Kroll, and Conway, *Exploration and Science*, 197–198.

79. Clark, *Lady with a Spear*, 209.
80. Ibid., 4–6; Sylvia A. Earle, *Dive! My Adventures in the Deep Frontier* (Washington, D.C.: National Geographic Society, 1999); and Bruce Greenlaugh, Navy SEAL diver, January 27, 2007, conversation with the author.
81. Valle and others, *Skin and Scuba Diving*, 8.
82. Hanauer, *Diving Pioneers*, 71.
83. Dugan, *Man Under the Sea*, 214–215.
84. Earle and Giddings, *Exploring the Deep*, 168.
85. Norton, *Stars Beneath the Sea*, 176–197; and Reidy, Kroll, and Conway, *Exploration and Science*, 191–193.
86. Pfister, "How to Build and Use a Diving Lung," 160; and Hanauer, *Diving Pioneers*, 12–13.
87. Valle and others, *Skin and Scuba Diving*, 8.
88. Joe Strykowski, *Diving for Fun: A Complete Textbook for Students, Instructors, and Advanced Divers* (1969; repr., Northfield, Ill.: Dacor Corporation, 1974), 1.
89. Valle and others, *Skin and Scuba Diving*, 11.
90. Ibid., 22. Ellipses in the original.
91. Willard Bascom and Roger Revelle, "Free-Diving: A New Exploratory Tool," *American Scientist* 41 (October 1953): 624–625.
92. Arthur C. Clarke, *The Challenge of the Sea* (New York: Holt, Rinehart and Winston, 1960).
93. "Sir Arthur C. Clarke: The Times Obituary," *Times (London)*, March 19, 2008, http://www.timesonline.co.uk/tol/comment/obituaries/article3582978.ece.
94. Arthur C. Clarke, *2001: A Space Odyssey* (New York: Signet, 1968); Clarke, *Challenge of the Sea*; and Clarke, *Voice Across the Sea* (New York: HarperCollins, 1975).
95. Clarke, *Boy Beneath the Sea*, photographs by Mike Wilson (New York: Harper and Row, 1958).
96. Clarke, *The Deep Range* (New York: Harcourt, Brace, 1957); and Clarke, *Dolphin Island: A Story of the People of the Sea*, (New York: Holt, Rinehart and Winston, 1963). In these years Clarke also published what became known as his Blue Planet Trilogy: *The Coast of Coral* (New York: Harper, 1957); *The Reefs of Taprobane: Underwater Adventures Around Ceylon* (New York: Harper, 1957); and *The Treasure of the Great Reef* (New York: Harper and Row, 1964).
97. Wernher von Braun, "Introduction," in Clarke, *Challenge of the Sea*, 7.
98. Dugan, *Man Under the Sea*, 215.
99. John Sweeney, *Skin Diving and Exploring Underwater* (New York: McGraw-Hill, 1955), vii–viii.
100. Robert B. Leamer, Wilfred H. Shaw, and Charles F. Ulrich, *Bottoms Up Cookery* (Gardena, Calif.: Mastergraphix, 1971).
101. Henrika Kuklick and Robert E. Kohler, eds., *Science in the Field, Osiris* 11 (1996).
102. Seabrook Hull, *The Bountiful Sea* (Englewood, N.J.: Prentice Hall, 1964), 221.
103. Gary M. Kroll, "Exploration in the Mare Incognita: Natural History and Conservation in Early Twentieth-Century America" (Ph.D. diss., University of Oklahoma, 2000); Reidy, Kroll, and Conway, *Exploration and Science*; Helen M. Rozwadowski, "Engineering, Imagination, and Industry: Scripps Island and Dreams for Ocean Science in the 1960s," in Rozwadowski and van Keuren, *Machine in Neptune's Garden*, 315–354.

Eight

James Rodger Fleming

Planetary-Scale Fieldwork

Harry Wexler on the Possibilities of Ozone Depletion and Climate Control

"The subject of weather and climate control is now becoming respectable to talk about." So began Harry Wexler's speech "On the Possibilities of Climate Control," given in early 1962 in Boston, Hartford, and Los Angeles.[1] Wexler, who studied meteorology at MIT and directed the office of meteorological research at the U.S. Weather Bureau, supported his claim by citing President John F. Kennedy's recent speech at the United Nations proposing "cooperative efforts between all nations in weather prediction and eventually in weather control."[2] Soviet Premier Nikita Khrushchev, flush with the success of two space spectaculars carrying Russian cosmonauts into orbit, had also mentioned weather control in his report to the Supreme Soviet in July 1961. Wexler noted that the subject had recently received serious attention from the President's Science Advisory Committee, the National Academy of Sciences Committee on Atmospheric Sciences, and the State Department. The Academy had recommended increased funding and the creation of a National Center for Atmospheric Research. The United Nations, with Wexler's scientific input through State, had recently recommended measures to advance the state of atmospheric science and technology in outer space "to provide greater knowledge of basic physical forces affecting climate and the possibility of large-scale weather modification."[3]

Wexler lectured in 1962 to the Boston Chapter of the American Meteorological Society, the Travelers Insurance Research Corporation in Hartford, and the UCLA Department of Meteorology. He assured his largely technical audiences that he was concerned not with the long and checkered history of cloud modification leading to more-or-less localized precipitation influences, but with planetary-scale manipulation of the Earth's short-wave and long-wave radiation budget that would result in "rather large-scale effects on general circulation patterns in short or longer periods, even approaching that of climatic change."[4] These effects detailed below, included *increasing* world temperature by several degrees by detonating up to ten H-bombs in the Arctic Ocean; *decreasing* world temperature by launching powder into an equatorial orbit to shade the Earth and make it look somewhat like Saturn and its rings; *warming* the lower atmosphere and *cooling* the stratosphere by artificial injections of water vapor or other substances; and, notably, *destroying all stratospheric ozone* above the Arctic Circle using a relatively small amount of a catalytic agent such as chlorine or bromine. Wexler was interested in both inadvertent climatic effects, such as might be created by rocket exhaust gases or other pollution, and purposeful effects, whether peaceful or hostile.

Today, techniques for cooling the Earth are very much in the news, especially in the wake of recent conferences sponsored by NASA and Harvard University on "Managing Solar Radiation" as a technological response to global warming.[5] The stratospheric ozone story is also very significant, given that the received history of the ozone-depleting chemical reactions dates only to about 1970 and certainly does not include Harry Wexler's role. Recently, I have been in personal communication with three notable ozone scientists about Wexler's early work: Nobel laureates in chemistry F. Sherwood Rowland and Paul Crutzen, and president of the National Academy of Sciences Ralph Cicerone. They agree that if Wexler had published his lectures in 1962, ozone-depletion chemistry and possibly diplomacy could have started more than a decade before it did.

First let's get some background with a long historical excursion on how to do planetary fieldwork, then look at Wexler's career, and finally examine inadvertent and purposeful climate control circa 1962.

By What Authority?

What do climate scientists know about climate change and how do they know it? By what authority and by what historical pathways have they arrived at this knowledge? How have scientists established privileged positions on phenomena that cover the entire globe, that have both natural and anthropogenic components, and that are constantly changing on a multiplicity of temporal and spatial scales? By what means and by what authority do the geo-engineers (or geo-scientific speculators) plan to manipulate or "control" the climate in response to technical or societal concerns? As we move into such uncharted territory, what lessons are applicable from the history of smaller-scale field studies? In short, what does a study of the long history of scientific practices in knowing and manipulating global climate have to offer to current pressing issues in science and public policy?[6]

Awareness

How have climate scientists gained awareness and understanding of phenomena that cover the entire globe and that are constantly changing on time scales ranging from geological eras and centuries to decades, years, and seasons? How was this accomplished by individuals immersed in and surrounded by the phenomena? How were privileged positions created and defined? The answers span at least three centuries and provide historical grounding for knowledge claims made about something as nebulous as global climate change. One approach, popular in the eighteenth century yet still with us today, was through appeals to authority. Then it was based on references to historical literature, first impressions of explorers, or the memories of the elderly; now it is based on having access to some unique global platform, most likely through space or computer technology, and by forging consensus through deliberative process such as the Intergovernmental Panel on Climate Change.

In the eighteenth century, however, personal authority was the rhetorical strategy of the Enlightenment and early American writers who wanted to support a particular climate-based theory of cultural development or decline. An early modern example of this derives from the work of Jean-Baptiste Abbé Du Bos (1670–1742), perpetual secretary of the French Academy, who linked cultural and climatic changes in his 1719 book *Réflexions critiques sur la poësie et sur la peinture*. Du Bos argued that artistic genius flourished only in countries with suitable climates (always between twenty-five and fifty-two degrees north), that changes in climate must have occurred to account for the rise and decline of the creative spirit in particular nations, and that the climate of Europe and the Mediterranean area was now warmer than it had been in ancient times. His idea that climate affected culture has an ancient heritage, traceable to Aristotle's *Politics*. But Du Bos added a temporal dimension to climatic determinism, linking climate *change* to the rise (and fall) of creativity in such eras as Greece in the third century BCE, imperial Rome under the Caesars, Renaissance Italy, and his own age: seventeenth-century France under the Sun King, Louis XIV. More-proximate sources of Du Bos's idea linking climatic and cultural change included the works of Jean Bodin, John Barclay, and Sir John Chardin. Du Bos in turn influenced other famous authors who invoked climate change, including Edward Gibbon, David Hume, and the Baron Montesquieu.

The ideas of Du Bos and his followers dominated climate discourse in the second half of the eighteenth century. Theirs was a powerful vision of climates shaping the course of empire and the arts, as well as a much more moderate European climate than in ancient times, largely due to the concerted activities of innumerable individuals engaged in deforestation, swamp drainage, and agriculture. Working with these ideas, Thomas Jefferson (1743–1826) held that the cold, moist American climate was undergoing even more rapid and dramatic amelioration caused by settlement, changes that would soon bring it into line with European norms. As I stated above, all this was based on appeals to authority. Such opinions were parroted by no less of an authority than the prominent nineteenth-century geologist Charles Lyell, who maintained that, "in the United States of North America, it is *unquestionable* that the rapid clearing of the country has rendered the winters less severe, and the summers less hot." We will return to issues of authority.

Data

Yet something new was beginning around this time. Jefferson was a strong advocate of collecting massive amounts of meteorological data over large areas and extended time periods in the hope of deducing climatic patterns and changes: "Measurements of the American climate should begin immediately, before the climate has changed too drastically. These measurements should be repeated. . . . once or twice in a century, to show the effect of clearing and culture towards the changes of climate."[7] Beginning in earnest in the 1780s, notably in Bavaria, networks of cooperative observers dutifully tended their instruments and gradually extended the meteorological frontiers, while scientists tabulated, charted, mapped, and analyzed the observations to provide climatic inscriptions. This process profoundly changed climate discourse and established the foundations of the science of climatology.

Although the examples are many, there is space here to mention but a few. In 1842, army physician Samuel Forry (1811–1844) published an analysis of weather data gathered by the Army Medical Department since 1814 at over sixty locations. Forry drew three main conclusions from his study: climates are stable, and no accurate thermometric observations indicate systematic climatic change; climates *are* susceptible to melioration by the changes wrought by the labors of man; but these effects are much less influential than those of latitude, elevation, and proximity to bodies of water. These results were in basic agreement with those expressed in Alexander von Humboldt's popular work *Views of Nature* (1808).

Other early climatologists extended the practice of privileging physical geography over human influence and of citing meteorological records rather than ancient authorities or the memories of the elderly. For example, Lorin Blodget, who conducted meteorological research with the Smithsonian

Institution and the U.S. Army, included a chapter in his *Climatology of the United States* (1857) titled "Permanence of the Principal Conditions of Climate." Other studies by prominent meteorologists Elias Loomis, Charles Anthony Schott, and William Ferrel arrived at a similar conclusion, arguing *against* the notion of a changing climate, at least on decades-to-centuries timescales. In 1889 Cleveland Abbe (1838–1916), chief scientist for the U.S. National Weather Service, wrote that the old debates about climate change had finally been settled. Abbe defined the climate as "the average about which the temporary conditions permanently oscillate; it assumes and implies permanence." Alluding to the recent discovery of the ice ages, Abbe conceded that:

> great changes have taken place during geological ages perhaps 50,000 years distant; but no important climatic change has yet been demonstrated since human history began. . . . It will be seen that rational climatology gives no basis for the much-talked-of influence upon the climate of a country produced by the growth or destruction of forests, the building of railroads or telegraphs, and the cultivation of crops over a wide extent of prairie. Any opinion as to the meteorological effects of man's activity must be based either upon the records of observations or on *à priori* theoretical reasoning. . . . The true problem for the climatologist to settle during the present century is not whether the climate has lately changed, but what our present climate is, what its well-defined features are, and how they can be most clearly expressed in numbers.[8]

Thus a major shift had occurred, from literary to empirical studies of climate. It is important to remember, however, that climate theorists of the time were focused on explaining the cause of ice ages, and that theories of human influence and climatic determinism would soon reappear—as the streams of data grew into overwhelming torrents.

Theory

The debate over climate change caused by human activities was winding down just about the time that scientists discovered that the Earth had experienced ice ages and interglacials—tremendous advances and retreats of the glaciers over geologic time periods. These discoveries, especially the need to explain multiple glaciations, produced a plethora of complex but highly speculative theories of climatic change involving astronomical, physical, mathematical, and geological evidence and principles. Prominent scientists such as Joseph Fourier, Joseph Adhémar, John Tyndall, James Croll, Svante Arrhenius, T. C. Chamberlin, and many others—all from different eras, and all in their own ways—sought to establish, from first principles, what the climate ought to be and how it ought to change.[9] Theoretical approaches focused not on weather data, but on the overall behavior of the oceans, the Earth's orbit, solar insolation, and the global carbon budget. Such explanations tended to be most satisfying to those scientists working within a particular disciplinary perspective. Most scientists had one favorite causal mechanism and only grudgingly admitted other possible secondary causes of climate change.

For example, in the mid-nineteenth century the Scottish geologist James Croll (1821–1890) was pioneering the theory that the Earth's climate "revolves" around the sun though the slow but methodical variation of its orbital elements—tilt, wobble, and eccentricity—triggering complex feedbacks in solar insolation, cloudiness, ocean currents, and ice sheets—a field we now call "climate dynamics."[10] The work of Croll's successor, Serbian engineer and mathematician Milutin Milankovitch now form (after more than a half century of contentious debate) the bedrock of theories of natural climate variability. Milankovitch cycles are widely considered to be the "pacemaker" of ice age and interglacial cycles and "trigger" of dynamic climate feedbacks.

Experiment

In the nineteenth century, radiant heat (or infrared radiation) became a focus of inquiry. John Tyndall (1820–1893) conducted experiments at the Royal Institution of Great Britain on the radiative properties of various gases, demonstrating that "perfectly colorless and invisible gases and vapours" were able to absorb and emit radiant heat.[11] He identified the importance of atmospheric trace constituents as efficient absorbers of long-wave radiation and as important factors in climatic control. Specifically, he established beyond a doubt that the radiative properties of aqueous vapor, carbonic acid, and ozone were of importance in explaining meteorological phenomena such as the formation of dew, the energy of the solar spectrum, and possibly the variation of climates over geological time. Among his most striking discoveries were the vast differences in the abilities of gases to absorb and transmit radiant heat. The "elementary gases," oxygen, nitrogen, and hydrogen, were almost transparent to radiant heat, while more complex molecules, even in very small quantities, absorb much more strongly than the atmosphere itself.

Concerning climate, Tyndall thought that changes in the amount of any of the radiatively active constituents of the atmosphere—water vapor, carbon dioxide, ozone, or hydrocarbons—could have produced "*all the mutations of climate which the researches of geologists reveal* . . . they constitute true causes, the *extent* alone of the operation remaining doubtful."[12] He gave credit to his predecessors for the intuition that "the rays from the sun and fixed stars could reach the earth through the atmosphere more easily than the rays emanating from the earth could get back into space." The experimental verification of this phenomenon, however, belonged to Tyndall. His carefully executed laboratory experiments clearly demonstrated that trace atmospheric constituents were active absorbers of heat radiation. His meteorological and climatological speculations kept alive what was called the "hothouse theory," and suggested to Svante Arrhenius, T. C. Chamberlin, and others that the Earth's heat budget may be controlled by changes in the trace constituents of the atmosphere.[13]

Models

It was the Swedish Nobel laureate in chemistry Svante Arrhenius (1859–1927), however, who connected the geophysical and the anthropogenic theories of climate change. In 1896, following Tyndall's earlier suggestion, Arrhenius demonstrated that variations of atmospheric CO_2 concentration could have a very great effect on the overall heat budget and surface temperature of the planet, and might trigger feedback phenomena that could account for glacial advances and retreats. Arrhenius's model relied heavily on the experimental and observational work of others, most notably Arvid Högbom's carbon cycle, in which volcanic eruptions were the chief source of carbon dioxide in the Earth's atmosphere. Note that Arrhenius did not write his 1896 essay because of any great concern for increasing levels of CO_2 caused by the burning of fossil fuels; instead, he was attempting to explain temperature changes at high latitudes that could account for the onset of ice ages and interglacials.

Human Influence

The notion that humans can influence climate, examined earlier for the case of the Enlightenment, reappeared in a new form in 1899 when Nils Ekholm, an associate of Arrhenius, wrote about "the influence of Man on climate." Ekholm pointed out that over the course of a millennium the accumulation in the atmosphere of carbonic acid from the burning of pit coal will "undoubtedly cause a very obvious rise of the mean temperature of the Earth." He also thought this effect could be accelerated by the "digging of deep fountains pouring out carbonic acid," or perhaps decreased "by protecting the weathering layers of silicates from the influence of the air and by ruling the growth of plants." By such means Ekholm pointed to the grand possibility that it might someday be possible "efficaciously to regulate the future climate of the Earth and consequently prevent the arrival of a new Ice Age."[14]

Arrhenius popularized Ekholm's observation in his book *Worlds in the Making*, noting that "the slight percentage of carbonic acid in the atmosphere may by the advances of industry be changed to a noticeable degree in the course of a few centuries." Arrhenius considered it likely that in future geological ages the Earth would be "visited by a new ice period that will drive us from our temperate countries into the hotter climates of Africa."[15] On the timescale of hundreds to thousands of years, however, Arrhenius speculated on a "virtuous circle" in which the burning of fossil fuels could help prevent a rapid return to the conditions of an ice age, and could perhaps inaugurate a new carboniferous age of enormous plant growth. Ekholm concurred, adding his own speculations about the possibility of deliberate climate modification to delay the onset of an ice age by burning coal exposed in shallow seams.

Yet in the early decades of the twentieth century, the carbon dioxide theory of climate change had fallen out of favor with most scientists. The dominant opinion was that, at current atmospheric concentrations, carbon dioxide already absorbed all the available long-wave radiation; thus any increases would not change the radiative heat balance of the planet but might augment plant growth. Other physical mechanisms of climatic change, although highly speculative, were given more credence, especially changes in solar luminosity, atmospheric transparency, and the Earth's orbital elements. The scientist responsible for reviving the carbon dioxide theory, and in this sense also the theory of human influence, was Guy Stewart Callendar (1898–1964). In 1938, Callendar reformulated the carbon dioxide theory by arguing that rising global temperatures and increased fossil fuel burning were closely linked. There is no space here to discuss the anthropogenic greenhouse effect in further detail. Suffice it to say that Harry Wexler mentioned the Callendar Effect in his 1962 lectures as one way we already inadvertently modify the global climate.[16]

Technology

The 1950s marked the dawning of the digital era of computer modeling of the general circulation and satellite monitoring of both weather patterns and the Earth's heat budget. This era also spawned a new breed of climate physicians and surgeons. The technological approach began long before that with the invention, distribution, and subsequent networking of the basic meteorological instruments, but it was magnified to the planetary level by the end of the 1950s. With the invention and standardization of meteorological instruments, the networking of meteorological observers, and the development of statistical analysis, a picture (albeit abstract and imperfect) of the climatic aspects of locations, regions, and even the globe emerged in the second half of the nineteenth century. Let's call it the "black-and-white marble" as illustrated by the work of James Henry Coffin's *Winds of the Globe* and William Ferrel's theoretical model of the general circulation. Everyone already knew that the clouds were typically white (or perhaps gray) and the sky was usually blue (at least on clear days before sunrise and sunset). The work of Mie and Rayleigh told the physicists why. This fuzzy image of the Earth's climate has been rendered three-dimensional in the twentieth century by the development of upper-air observations, extended into the indefinite past by paleoclimatic techniques, and, finally, globalized in the era of satellite remote sensing. It is important to remember that there were major technological breakthroughs relevant to meteorology in the 1840s, linked to the speed of communication and connectivity of the telegraph; the 1910s, as humans took to the air; and the 1930s, as radio stations broadcast the weather and miniaturized sensors flew to the top of the troposphere on balloon-borne radiosonde probes. Many scientists today are working on links between satellite remote sensing and more sophisticated weather and climate models. They are hoping, through advances in technology, to provide new privileged positions. For most scientists the goal is better understanding of climate; for some it is also climate prediction, although such a crystal ball might severely disrupt commerce and international affairs; for a fringe group, it is climate control, which, as John von Neumann warned as early as 1956, may be more dangerous than nuclear proliferation.

In the 1950s, with the advent of digital computer models it became possible to solve and display graphically solutions to the equations of atmospheric motion. So-called general circulation models trace their origins to the work of MIT meteorologist Norman Phillips, who focused on emulating the dynamics of atmospheric flows. In the decades that followed, Joseph Smagorinsky led a team at NOAA's Geophysical Fluid Dynamics Lab that extended this work to diagnostic climate models that incorporated additional complexity. "Model experiments" have continued ever since, combining technology and massive amounts of data, and have provided the most widely used and authoritative assessments of climate sensitivity and projected change.

Authority, data, theory, experiments, models, technology, authority . . . a braided narrative leading from the eighteenth to the twenty-first centuries, terminating today, with ends still frayed, with integrated assessments of human influence, human impacts, and layer on layer of recommendations, although as yet with insufficient actions.

The Importance of Being Harry Wexler

Harry Wexler, one of the most influential meteorologists of the first half of the twentieth century and a pioneer in interdisciplinary atmospheric science, was born in 1911 in Fall River, Massachusetts, and died in August 1962 at age fifty-one unexpectedly (figure 8.1).[17] A 1932 Harvard graduate in mathematics, Wexler was mentored at MIT by the noted meteorologists Carl-Gustav Rossby, Hurd C. Willett, and Bernhard Haurwitz. Wexler held research and teaching positions in meteorology at MIT, the U.S. Weather Bureau, the

Figure 8.1

Harry Wexler (1911–1962). *From the Harry Wexler Papers, Library of Congress.*

University of Chicago, and the U.S. Air Force. Following his honorable discharge from the military in January 1946 with the rank of lieutenant colonel, Wexler returned to the weather bureau, becoming chief of the Special Scientific Services division. As head of research, Wexler encouraged the development of new technologies, including airborne observations of hurricanes, sounding rockets, weather radar, nuclear tracers, the use of electronic computers for numerical weather prediction and general circulation modeling, and satellite meteorology.

Wexler served on numerous panels and committees, including the military's Research and Development Board and the National Advisory Committee for Aeronautics (NACA) Subcommittee on Meteorological Problems. He took many trips to Princeton as liaison to the Institute for Advanced Study's meteorology program, which was attempting to develop a numerical forecast model using a digital computer. He was a delegate to the Toronto meetings of the International Meteorological Organization, where he served with geophysicist Sydney Chapman on the International Ozone Commission and chaired the U.S. delegation on aerology. He was a member of the Advisory Committee on Reactor Safeguards for the Atomic Energy Commission and served as a U.S.

delegate to the "Atoms-for-Peace" Conference in Geneva in 1955. He was the vice president of the American Meteorological Society, and the chairman of the Upper-Atmosphere Committee of the American Geophysical Union. He also chaired the NACA Special Committee for the Upper Atmosphere; the Geophysical Research Panel of the Scientific Advisory Board to the Chief of Staff, U.S. Air Force; and a study group on Meteorological Aspects of the Effects of Atomic Radiation for the National Academy of Science. He served on the National Research Council Space Science Board as well as its Committee on Arctic and Antarctic Research. In all these roles he was at the cutting edge of atmospheric research and its social applications. His interests were in global processes (including a study he commissioned on the atmospheres of other planets), and he contributed to a new global focus for the atmospheric sciences.

Wexler was closely associated with the Weather Bureau Research Station, Mauna Loa Observatory. The observatory staff directly reported to Wexler, and he gave it a special measure of his interest and attention. He visited the observatory often and took intense pleasure in the ecology of the lush rain forest. A number of atmospheric base line and other measurements were conducted there, including, significantly, the ongoing series of global background measurements of carbon dioxide begun by Scripps Institution of Oceanography chemist Charles David Keeling in 1958.

During the International Geophysical Year (IGY) of 1957–58, Wexler served in a number of positions for the U.S. National Committee, including as a member of the Technical Panel on Meteorology; deputy to chief of the Weather Bureau, F. W. Reichelderfer, on the IGY Executive Committee; consultant to the Antarctic Committee; member of the ad hoc Arctic and Equatorial Committee; and member of the ad hoc Panel for Radioactivity of the Air. Most notably, he was the chief scientist of the USNC/IGY Antarctic Programs.[18] In 1957, Wexler wrote a classic paper on "Meteorology in the International Geophysical Year" that highlighted some of the fundamental issues in the understanding of the atmosphere, meteorology's relationship to other geophysical sciences, and the importance of Antarctic science, climate science, and weather satellites. He spearheaded ozone measurements globally during the IGY with special focus on Antarctica. Wexler maintained his membership in the International Ozone Commission through the next three meetings in Oxford (1959), Helsinki (1960), and Arosa, Switzerland (1961).

Wexler was a pioneer, spokesman, and promoter of the use of satellites in meteorology (figure 8.2) and was actively involved in the Joint Meteorological Satellite Advisory Committee in 1959, where he commented at the First National Conference on Peaceful Uses of Space. He believed that information

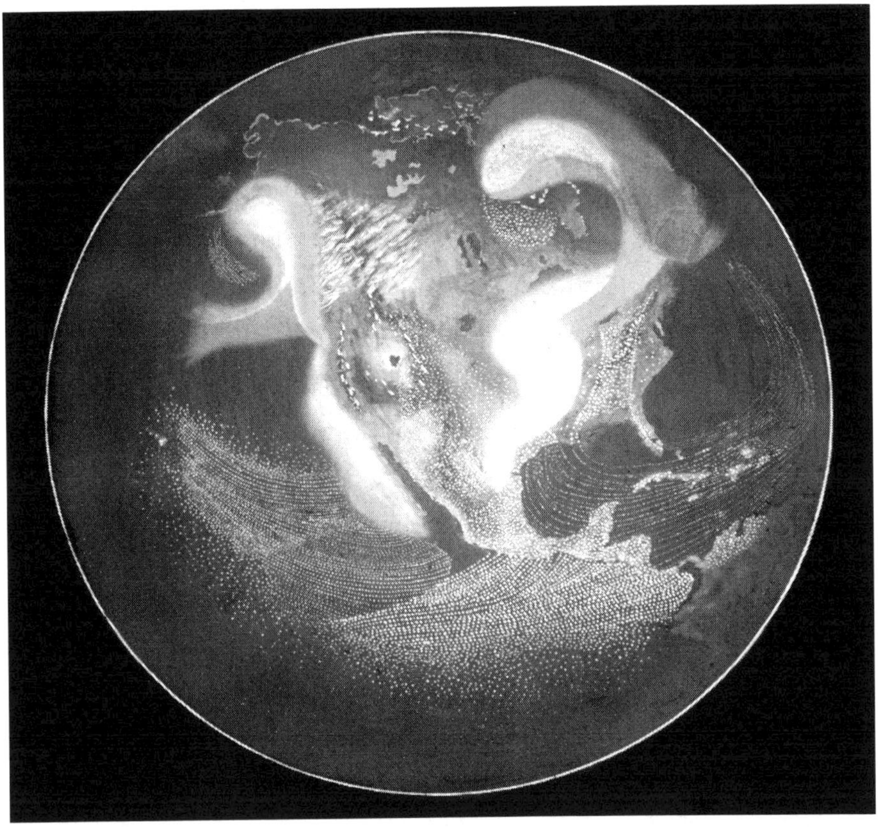

Figure 8.2

Weather systems over North America as they might appear from a satellite four thousand miles above Amarillo, Texas, on June 21. Harry Wexler, director of meteorological research, U.S. Weather Bureau, commissioned the painting in 1954. Surface features are drawn taking into account the Earth's normal colors, reflectivity of sunlight, and scattering and depleting effects of light passing through the atmosphere, with calculated brightness of various cloud types. Weather features include a family of three cyclonic storms extending southwest from Hudson Bay to Texas; a similar system over the Gulf of Alaska; a small hurricane developing near Puerto Rico; a meeting zone of northeast and southeast trade winds, extending west of the Isthmus of Panama to mid-Pacific; a line squall in the eastern United States; scattered cumulus clouds over heated land areas; lenticular clouds usually found where the jet stream crosses mountains, as over the northern Canadian Rockies; and low stratus and fog off the California coast, over the Great Lakes, and in the Newfoundland area. For a full-color image see James R. Fleming, "A 1954 Color Painting of Weather Systems as Viewed from a Future Satellite," *Bulletin of the American Meteorological Society* 88 (2007): 1525–1527.

Image from the Harry Wexler Papers, Library of Congress; original painting hangs in the conference room of the National Environmental Satellite, Data, and Information Service, Silver Spring, Maryland.

gathered from satellites on the international scale would be of great value to everyone in the world for warning of severe weather and other meteorological changes. In 1961 and 1962 he served as the lead negotiator for the United States in talks with the Soviet Union concerning the joint use of meteorological satellites in the World Weather Watch. Earth radiation heat budget experiments from satellite were of particular interest to him.

Wexler on the Possibilities of Ozone Depletion and Climate Control

In his speeches on climate control in 1962,[19] after assuring his audience of the respectability of the subject and alluding to the recent remarks of Kennedy and Khrushchev, Wexler discussed recent developments in society, including increasing pollution both from industry and from rockets, and recent developments in science, including computing and satellites, that led him to believe that manipulating and controlling large-scale phenomena in the atmosphere were distinct possibilities. Concerning the pollution, Wexler mentioned carbon dioxide emissions as an example, and cited a 1961 study by the Geophysics Corporation of America on modification of the Earth's upper atmosphere by rockets. Wexler thought that atmospheric scientists were engaged in what he called the "natural evolutionary growth" of a scientific discipline, from understanding to prediction, and ultimately to control—control even at large-scale weather and global climate.

According to Wexler, this concept and philosophical approach was stressed at the birth of modern computing, beginning in October 1945, when the noted inventor and electrical engineer Dr. V. K. Zworykin, then at the RCA Laboratories, Princeton, New Jersey, wrote his now all-but forgotten, mimeographed "Outline of Weather Proposal."[20] Zworykin imagined that a perfectly accurate machine would predict the immediate future state of the atmosphere and identify the precise time and location of leverage points or sensitive conditions so that a paramilitary, rapid deployment force might be sent out *into the field* to intervene in the weather as it happens—literally to pour oil on troubled ocean waters or even set fires or detonate bombs to disrupt storms before they formed, deflect them from populated areas, and otherwise control the weather. Zworykin ended the main body of his proposal with the following statement:

> The eventual goal to be attained is the international organization of means to study weather phenomena as global phenomena and to channel the world's weather, as far as possible, in such a way as to minimize the damage from catastrophic disturbances, and otherwise to benefit the world to the greatest extent by improved climatic conditions where possible.

> *Such an international organization may contribute to world peace by integrating the world interest in a common problem and turning scientific energy to peaceful pursuits. It is conceivable that eventual far-reaching beneficial effects on the world economy may contribute to the cause of peace.*[21]

John von Neumann, the multi-talented mathematician extraordinaire at the Institute for Advanced Study in Princeton, endorsed Zworykin's view, writing to him, "I agree with you completely. . . . This would provide a basis for scientific approach[es] to influencing the weather." Using computer-generated predictions, von Neumann wrote, weather and climate systems "could be controlled, or at least directed, by the release of perfectly practical amounts of energy," or by "altering the absorption and reflection properties of the ground or the sea or the atmosphere." It was a project that neatly fit von Neumann's overall philosophy: "All stable processes we shall predict. All unstable processes we shall control."[22] Zworykin's proposal was also endorsed by the oceanographer Athelstan Spilhaus, then a U.S. Army major, who ended his letter of November 6, 1945, with these words: "In weather control, meteorology has a new goal worthy of its greatest efforts."[23]

Wexler continued his introductory remarks: "So with the endorsement by statesmen and responsible scientists alike, there is no question about the respectability of this subject—but what about its attainability? Even in this day of global experiments, such as the world-wide ARGUS electron seeding of the Earth's magnetic field at 300 miles height,[24] man and machinery orbiting the Earth at 100 miles seventeen times in one day, and 100 megaton bombs—are we any closer to some idea of the approaches which could lead to an eventual 'solution' [to the problem of climate control]?" He noted "a growing anxiety" in the public pronouncements that "man, in applying his growing energies and facilities against the power of the winds and storms, may do so with more enthusiasm than knowledge and so cause more harm than good."[25]

Warning his audience that he did not intend to cover all possibilities, "but just a few that have struck my fancy . . . *limited primarily to interferences with the Earth's radiative balance on a rather large scale* [original emphasis] . . . I shall discuss in a purely hypothetical framework those atmospheric influences that man might attempt deliberately to exert and also those which he may now be performing or will soon be performing, perhaps in ignorance of its consequences. We are in weather control *now* whether we know it or not." He continued, "We have for decades been releasing huge quantities of carbon dioxide and other gases and particles to the lower atmosphere. It is recognized that this atmospheric pollution may have serious effect not only on health but on global radiation or heat balance which is the cause of our present pattern of climate and weather."[26]

At this point in his lecture, Wexler showed some twenty technical slides of the atmosphere's radiative heat budget and discussed means of manipulating it. He concluded with a grand summary of various techniques, including the following:

(a) *increasing* global temperature by 1.7°C by injecting ice crystals into the polar atmosphere; Wexler compared the Soviet proposal to dam the Bering Straits with his hypothetical method of detonating ten H-bombs in the Arctic Ocean;[27]
(b) *lowering* global temperature by 1.2°C by launching a ring of dust particles into equatorial orbit, a modification of a Soviet proposal to warm the Arctic;
(c) *lowering* the tropopause and *cooling* the stratosphere by injecting water vapor into the stratosphere, thus modifying the infrared radiation budget (as in the so-called Highwater Experiment); and
(d) *destroying all stratospheric ozone,* raising the tropopause, and cooling the stratosphere by an injection of a catalytic deozonizing agent such as chlorine, bromine, or perhaps fluorine.

On the latter topic, Wexler turned to Sydney Chapman's 1934 presidential address to the Royal Meteorological Society, which asked, "Can a hole be made in the ozone layer?" That is, can all or most of the ozone be removed from the column of air above some chosen ground area. Chapman was thinking of a temporary and localized event, somewhat like a solar eclipse, that would provide a window for astronomers to extend their observations some hundreds of ångstroms further into the ultraviolet without the interference of atmospheric ozone. Possible health effects of human exposure to short-wave radiation did not appear to Chapman to be an important issue, since the hole he was contemplating would be localized, probably in a remote area, and would be short-lived, somewhere between a day and an hour, and timed for the benefit of astronomers only. Cutting such a hole, Chapman continued, would require "the discharge of a deozonizing agent," perhaps by airplanes, balloons, or rockets. Two possibilities came to mind: a large amount of a one-to-one destructive agent, such as hydrogen, that would reduce O_3 molecules to O_2, or "some *catalyst* which, without itself undergoing permanent change, could promote the reduction of large numbers of ozone molecules in succession." Although the choice of the agent would have to be left to the chemists, Chapman concluded that "the project of making a [temporary] hole in the ozone layer does not seem quite impossible of achievement."[28]

In an internal memo to Weather Bureau colleagues in 1961, Wexler wrote: "In the QJRMS for 1934 Sydney Chapman proposed making a temporary 'hole' in the ozone layer by inserting a substance which could be oxidized by the ozone. He suggested that hydrogen might be dispersed but wondered if there might be a catalyst gas or fine powder which might perhaps be dispersed in smaller quantities than the 1 to 1 ratio hydrogen would require. . . . Could you or your colleagues suggest suitable agents that might do the job with maximum efficiency consistent with the least weight?"[29] Wexler lectured on this at a Weather Bureau seminar and received the following note from his colleague Bill Malkin:

> After leaving the seminar today, at coffee break time, I was challenged by your interest in Chapman's ozone hole suggestion, to try to recall where I had seen this or a similar reference. I was able to locate the reference I had seen. The reference to an "ozone hole" is marked on page 211 of the attached stuff.
>
> Incidentally, Dr. Wexler, perhaps you may wish to suggest, to our national defense research arm, that serious consideration be given to the possibility of artificially and temporarily altering (up or down) the ozone concentration over an area, as a most effective weapon.[30]

Seeking further advice on how to cut a "hole" in the ozone layer, Wexler turned to chemist Oliver Wulf at Caltech, who suggested that chlorine or bromine atoms might act in a catalytic cycle with atomic oxygen to destroy thousands of ozone molecules. In a handwritten note composed in January 1962, Wexler scrawled the following: "UV decomposes $O_3 \rightarrow O$ in presence of halogen like Br. $O \rightarrow O_2$ recombines and so prevents more O_3 from forming. 100,000 tons Br could theoret[ically] prevent all O_3 north of 65°N from forming." Elsewhere Wexler noted: "$Br_2 \rightarrow 2$ Br in sunlight destroys $O_3 \rightarrow O_2 +$ BrO." This is essentially the basis of the modern ozone-depleting chemical reactions. Wexler concluded that he was not making proposals to intervene, but was involved in studying the basic equations and engineering aspects of general circulation research, including the natural behavior of the atmosphere, unintentional effects, and aspects of particular interest to the Department of Defense. Wexler was concerned that inadvertent damage might be done to the thin layer of ozone by increased pollution from rocket exhaust or by near-space experiments gone awry. Chapman's 1934 speculation about purposefully cutting a temporary "hole in the ozone layer" for the benefit of astronomers paled in comparison to possible military interest in waging geophysical warfare by attacking the ozone layer over a rival nation.

The Climate Engineers

In 1965, three years after Wexler's death, President Johnson's Science Advisory Committee issued a report called *Restoring the Quality of Our*

Environment. After estimating the future increase of anthropogenic CO_2 due to fossil fuel burning and its likely negative impact on climate, the report suggested that geo-engineering options, or as they put it, "the possibilities of deliberately bringing about countervailing climatic changes, . . . need to be thoroughly explored."[31] As an illustration, they pointed out that the earth's albedo could be increased by 1 percent by dispersing buoyant reflective particles on the sea surface at an annual cost, not considered excessive, of about five hundred million dollars. Reducing fossil fuel use was not mentioned as an option.

In 1977 Cesare Marchetti tackled the problem of CO_2 control in the atmosphere by proposing a kind of "fuel cycle," to collect and inject it, for example, into the Mediterranean at the Straits of Gibraltar. About the same time, the Soviet scholar M. I. Budyko emphasized modification of the aerosol layer of the stratosphere. Technical proposals continued to dominate. A 1992 National Academy of Sciences report, *Policy Implications of Greenhouse Warming*, advised that the United States should conduct research in schemes to cool the Earth if global warming gets out of hand. Proposals included orbiting a fleet of space mirrors or spraying sulfur dioxide into the stratosphere to reflect solar radiation back to space; turning the oceans into soupy, green algae blooms to sequester excess carbon; or setting up gigantic "soot generators" to shade the planet.

Recently, really only since about 2005 or so, leading figures in the climate modeling community have alluded to themselves as "planetary physicians" and have appropriated descriptive metaphors from medical practice. The Earth is running a "fever" of one degree compared to a century ago, and the prognosis is not good. The fever will likely worsen substantially in the century to come, leading to melting ice caps, rising sea level, massive disruptions in water supply, killer heat waves, and stronger hurricanes. Humans, by adding to the greenhouse effect, are implicated through lifestyle and excess, analogous to the health risks facing an overweight, heavy-drinking chain smoker. The "planetary physicians" recommend drastic changes—such as a strict carbon "diet" to stabilize the amount of carbon dioxide in our atmosphere, and other mitigation strategies such as renewable energy sources and energy efficiency to avoid the most dangerous climatic consequences. Compliance with this advice, however, will involve all of humankind and will likely necessitate complete restructuring of policies and even polities worldwide. An emerging breed of planetary "surgeons" thinks such voluntary compliance is highly unlikely and that more unilaternal and invasive techniques will be necessary.[32]

"Human-built volcanoes" injecting specially engineered reflective particles into the Arctic atmosphere using artillery guns, or spewing plumes of sulfates directly into the stratosphere, might artificially cool the planet, but at what cost, with what side effects, and, perhaps most important, by what authority? Some nations, historically, have desired an open Arctic Ocean, while others want to engineer the growth of ice sheets. Do we need another UN Environmental Modification Convention like the one ratified in 1978 to deal with environmental disputes caused by unilateral climate engineering? If a planetary thermostat is to be built, who will build it and where will it be located? In Geneva? At The Hague? Or at Lawrence Livermore National Laboratory?

On February 2, 2007, the Intergovernmental Panel on Climate Change reaffirmed the scientific consensus that it is "very likely" humanity is causing climate warming. There is no consensus, however, on the much more complex issue of what to do about it. Invoking the unlikelihood of carbon reductions being accomplished voluntarily, some prominent climate scientists, including Nobel laureate Paul Crutzen and others, are now suggesting that they can provide cheap, reliable "technological fixes" for the climate system through macro-engineering options that include albedo modification (turning the blue sky milky white), ocean fertilization (turning the blue Pacific soupy green), and other invasive techniques of "planetary surgery."

Currently there is a cacophony of scientific and popular opinion surrounding the issue of climate techno-fixes. Some "geo-scientific speculators" are supremely confident their techniques to change the Earth's albedo will work, cheaply and efficiently, with few adverse side effects. Others are more cautious, while actively seeking funding for climate control experimentation. Scientists, however, are notoriously ill suited, both by training and perspective, to analyze the historical, political, and ethical dimensions of such issues.

I was the sole historian at the NASA conference in November 2006 on "managing solar radiation," where I explained to the geo-engineers—many of whom consider themselves heroic pioneers, the first generation capable of alleviating or averting global natural disasters—that "history matters." I recounted the long and checkered history of the charlatans and sincere but deluded scientists and engineers who preceded them in the field of weather and climate control, reminding them they were not exempt from this history. I told them that when it comes to global ecology and the fate of the planet, "back-of-the-envelope calculations are not good enough"; neither are best climate model simulations. I told them if we fail to heed the lessons of history, and fail to bring history's perspectives to bear in thinking about public policy, we risk repeating the mistakes of the past, in a game with much higher stakes. One participant commented, as an aside, that this dismal history reminded him of the history of the eugenics movement. A follow-up meeting, held in Cambridge, Massachusetts, in November 2007, was notable for who was missing: everyone except technocratically oriented, white, middle-class males.

In facing the challenge of possible climate catastrophe, what shall we do? Doing nothing or too little is clearly wrong, but so is doing too much. Leaving these decisions to scientists and engineers is also clearly wrong. My goal is to open up the discussion to a broader audience and to see history, science, and policy as equal and dynamic partners in this discussion.

Conclusion

Framed in the context of both a broad survey of climate history and a quick look at modern climate-engineering concerns, it is clear that Harry Wexler met all the historical criteria for qualifying to speak authoritatively about the otherwise "nebulous" subjects of climate, climate change, and climate control. He was on all the scientific panels and advisory boards, had access to and helped collect global climate data, understood the theoretical issues and their complexity, and promoted and advanced the latest technologies, most relevantly general circulation modeling and satellite heat budget measurements.

In the summer of 1962 Wexler accepted an invitation from the University of Maryland Space Research and Technology Institute to lecture on "The Climate of Earth and Its Modifications," and might have, under normal circumstances, prepared his ideas for publication, as he had done in his 1958 article in *Science*, "On Modifying the Weather on a Large Scale."[33] But Wexler was cut down by a sudden heart attack on August 11, 1962, during a working vacation at Woods Hole, Massachusetts. The documents relating to his career, especially his remarkable work on ozone depletion and climate control, headed into the archives, probably not to be seen and certainly not to be reevaluated until today. Oliver Wulf apparently did not pursue this line of inquiry either, but his associate at Berkeley, the chemist Frederick Kaufman, remained involved in what eventually developed into the well-known and well-documented supersonic transport and ozone-depletion issues of the late 1960s and early 1970s. But the issue of a technological fix for climate—perhaps by intervening in the radiation budget of the planet (and thus its general circulation), perhaps by ocean engineering or some other approach, but undoubtedly through techniques of planetary fieldwork—remains very much alive.

Epilogue

This essay places global environmental history in conversation with the history of the field sciences, with the frame of reference at a planetary scale. It addresses the social and technological construction of authority as individuals and groups of scientists seek to convince others of their beneficence. Seeking to move both the Earth and its inhabitants on the nebulous topic of climatic change, they seek or construct Archimedean levers and "places to stand." When this essay was presented in Philadelphia in 2007, it stimulated extensive and active discussion. One aspect involved the huge economic and military interests represented by geo-engineering, and the contrast with smaller-scale environmental practices, such as academic fieldwork in a particular ecotone. Is geo-engineering unstoppable? Perhaps, unless historical analysis helps reveal its checkered past and counter-determinist aspects. Still, since studies of climate change history are as-yet relatively unknown, some in the group argued that whatever we do about climate change, it will "inevitably" involve a technological fix. Although humanists and social scientists may never be fully "in charge" of such decisions, it is clear that their voices need to be heard on issues as diverse as carbon taxes, emission reductions, and even space mirrors. A "middle path" (between doing too little and doing too much) needs to be defined, one that is open, democratic, international, and articulated by many voices and composed of many small-scale pieces. This led to the final set of questions: Should history be normative? Should it be used to persuade people about particular courses of action? Are issues of normativity particularly acute when it comes to global science issues? The alternative may be irrelevance. Voiced differently, should the profession of history be constrained to an ivory tower or should we, like all other scholars, define a public intellectual stance, especially when our research helps illuminate some of the pressing issues of our times. I think it is clear where I stand.[34]

Notes

1. Harry Wexler, "On the Possibilities of Climate Control," Boston Chapter of the American Meteorological Society, January 9, 1962; Travelers Insurance Research Corporation, Hartford, Conn., January 11, 1962; and UCLA Department of Meteorology, [as Regent's Lecturer in Meteorology], February 28, 1962, notes in Manuscript Division, Library of Congress, Wexler Papers, Box 18, Speeches and Lectures, 1962.
2. John F. Kennedy, Address in New York City Before the General Assembly of the United Nations, September 25, 1961. See also Kennedy's press conference of March 1, 1961. Transcripts available from John F. Kennedy Presidential Library and Museum, http://www. jfklibrary.org/.
3. United Nations General Assembly, "Resolution on the Peaceful Uses of Outer Space," December 20, 1961, http://www.unoosa.org/.
4. On weather control history see James R. Fleming "The Pathological History of Weather and Climate Modification: Three Cycles of Promise and Hype," *Historical Studies in the Physical Sciences* 37 (2006): 3–25.
5. James R. Fleming, "The Climate Engineers: Playing God to Save the Planet," *Wilson Quarterly* 31 (Spring 2007): 46–60; and Cornelia Dean, "Experts Discuss

Engineering Feats, Like Space Mirrors, to Slow Climate Change," *New York Times*, November 10, 2007.

6. The first section of this essay extends an argument first made in James Rodger Fleming, *Historical Perspectives on Climate Change* (New York: Oxford University Press, 1998).
7. Thomas Jefferson to Lewis E. Beck, 16 July 1824, in *The Writings of Thomas Jefferson*, ed. Albert Ellery Bergh (Washington, D.C.: Thomas Jefferson Memorial Association of the United States, 1907), 15:71–72.
8. Cleveland Abbe, "Is Our Climate Changing?" *Forum* 6 (1889): 678–688, on pp. 679, 687–688.
9. Original papers by Fourier, Tyndall, Arrhenius, and Chamberlin, with commentary, are reproduced in James R. Fleming, "Climate Change and Anthropogenic Greenhouse Warming: A Selection of Key Articles, with Annotations, from 1824 to 1995" (National Digital Science Library, 2008), http://wiki.nsdl.org/index.php/PALE:ClassicArticles/GlobalWarming.
10. James R. Fleming, "James Croll in Context: The Encounter Between Climate Dynamics and Geology in the Second Half of the Nineteenth Century," *History of Meteorology* 3 (2006): 43–53, http://www.meteohistory.org/.
11. Fleming, *Historical Perspectives*, 70.
12. John Tyndall, "On the Absorption and Radiation of Heat by Gases and Vapours, and on the Physical Connection of Radiation, Absorption, and Conduction," *Phil. Mag.*, ser. 4, 22 (1861): 276–277.
13. See Fleming, *Historical Perspectives.*
14. Nils Ekholm, "On the Variations of the Climate of the Geological and Historical Past and Their Causes," *Quarterly Journal of the Royal Meteorological Society* 27 (1901): 60–61; first published in Swedish in 1899 and reproduced in Fleming, "Climate Change and Anthropogenic Greenhouse Warming."
15. Fleming, *Historical Perspectives*, 82.
16. James Rodger Fleming, *The Callendar Effect: The Life and Work of Guy Stewart Callendar (1898–1964), the Scientist Who Established the Carbon Dioxide Theory of Climate Change* (Boston: AMS Books, 2007).
17. There is no secondary literature of note on Wexler, with the exception of Diane Belanger, *Operation Deepfreeze* (Boulder: University Press of Colorado, 2006); and Sepideh Yalda, "Harry Wexler," *Complete Dictionary of Scientific Biography* (Detroit: Thomson Gale, 2007), 25:273–276. The biographical sketch presented here was constructed exclusively from the Harry Wexler Papers, Manuscript Division, Library of Congress, which cover the years 1929 to 1962 and contain information relating to all the major areas of his career. A list of his published works appears in Malcolm Rigby and Pauline A. Keehn, "Bibliography of the Publications of Harry Wexler," *Monthly Weather Review* 91 (1963): 477–481.
18. James Rodger Fleming, "Making Meteorology Global: Polar Weather and Climate Research in the Career of Harry Wexler, 1933–1962," in *Globalizing Polar Science: Reconsidering the Social and Intellectual Implications of the International Polar and Geophysical Years*, ed. Roger D. Launius, James Rodger Fleming, and David H. DeVorkin (New York: Palgrave Studies in the History of Science and Technology, 2010).

19. Wexler, "On the Possibilities of Climate Control," draft, January 8, 1962, Wexler Papers, Box 18.
20. V. K. Zworykin, "Outline of Weather Proposal" (Princeton, N.J.: RCA Laboratories, 1945), copy in Wexler Papers, Box 18; reproduced in *History of Meteorology* 4 (2008): 57–78, http://meteohistory.org.
21. Ibid., 8; original emphasis.
22. Von Neumann to Zworykin, October 24, 1945, included in Zworykin, "Outline of Weather Proposal."
23. A. F. Spilhaus, "Comments on Weather Proposal," November 6, 1945, included in Zworykin, "Outline of Weather Proposal."
24. Other sorts of heavy-handed interventions have been attempted in the spirit of "seeding." In Project Argus, conducted in 1958 just six months after the discovery of the Van Allen radiation belt, the U.S. military and the Atomic Energy Commission detonated three atomic bombs 480 kilometers above the South Atlantic Ocean and two hydrogen bombs 160 kilometers over Johnston Island in the Pacific to seed the ionosphere with electrons. Another such seeding occurred in Project Highwater in 1962 when Saturn I test flights dumped eighty-six thousand kilograms of water at an altitude of 105 kilometers.
25. Wexler Papers, Box 18, "Further Justification for the General Circulation Research Request for FY 63," draft, February 9, 1962.
26. Wexler, "Further Justification."
27. Wexler, "Modifying Weather on a Large Scale," *Science*, n.s., 128 (1958): 1059–1063.
28. Sydney Chapman, "The Gases of the Atmosphere," *Quarterly Journal of the Royal Meteorological Society* 60 (1934): 127–142, on pp. 133–135.
29. "Deozonizer," memo from Wexler to Weather Bureau colleagues W. D. Komhyr, Gilbert Kinzer, and Sean Twomey, November 24, 1961, Box 13, Wexler Papers.
30. Bill Malkin to Wexler, November 22, 1961, Box 12, Wexler Papers.
31. President's Science Advisory Committee, *Restoring the Quality of Our Environment: Report of the Environmental Pollution Panel* (Washington, D.C.: Executive Office of the President, 1965), 111–133.
32. For details see Fleming, "Pathological History"; and Fleming, "Climate Engineers."
33. Harry Wexler, "Modifying Weather on a Large Scale," *Science*, n.s., 128 (1958): 1059–1063.
34. This is the theme of James Rodger Fleming, *Fixing the Sky: The Checkered History of Weather and Climate Control* (New York: Columbia University Press, 2010).

Nine

Robert E. Kohler

History of Field Science

Trends and Prospects

This volume of essays on the history of the field sciences—the (partial) record of an informal conference in May 2007—affords an occasion to reflect on the changes that have occurred in this subject since *Science in the Field*, the 1996 volume of *Osiris* edited by Henrika Kuklick and myself.[1] Have our predictions been born out; and what new turns has the history of field science taken in the intervening ten-plus years? And where does it seem to be heading in the next ten or so?

Kuklick and I had several purposes in assembling our volume in 1996. The first and most important was to make the field sciences more inviting and visible subjects for historians of science, dispelling their lingering second-class status vis-à-vis the laboratory sciences. A second was to move from a discipline-centered history of science toward a more ecumenical and comparative approach, by lumping in one category a set of sciences that have quite different subject matters but are alike in being carried out in the field. This categorical gambit, we hoped, would highlight the functional elements of place and practice—themes relevant to the history of all the sciences but pursued most advantageously in the field, where place always matters and practices are never routine.

Much has changed since 1996. Most strikingly, field science has become a respectable and even fashionable area for historical work. Studies of it, once rarely to be found in journals or at conventions of history of science societies, are now common. The growing number of recent dissertations is evidence that ambitious and able young scholars see the field sciences as areas of fresh opportunity in an increasingly well-occupied intellectual terrain. In addition, ideas of place and practice have become the common tools in the history of all the sciences, no longer needing special assistance from histories of field science, though these themes are still most strikingly clear in the case of field science.

With widespread acceptance have come other, unexpected changes. Historians seem less inclined now than formerly to assert the specialness of the field sciences as a group, and rarely if ever treat these comparatively. There is less interest now in their family resemblances and differences, or in generic differences between the sciences of the field and those of the lab. Some scholars deny that this categorical distinction is very useful or even that it exists at all—wrongly, in my view. In sum, the field sciences have become yet another subject for historical work, needing no favored status and no special pleading.

There is nothing wrong with this picture. It is how marginal fields of scholarship edge their way into mainstreams: by first asserting specialness, to attract talents, and then downplaying their specialness for a more assimilating identity. It's what being mainstream means: *not* being special—and thus by implication marginal—but doing what *everyone* does.

One wonders, though, if something vital may not be lost in assimilation. For many field scientists the categorical distinction between field and lab science is real enough, though sometimes annoying. If we historians cease to put our minds to what is special and particular about the field sciences, might we not also become less inclined to attempt big-picture history? Denying the force of the categorical distinction, might we not become less alert to the dynamic cultural processes that often occur across such categorical boundaries? Boundaries encourage innovation in science, as Tom Gieryn has argued and as I have tried to show in the case of field ecology.[2] Declare social categories non-existent and taboo, and the world of lived experience goes out of focus.

It is a peculiarity of the field sciences as a group that virtually every one straddles a spatial and social boundary—between lab and field, built and natural environments, and tabletop and outdoor practices, with their different conventions of establishing credibility and trust. In the case of the sciences that collect and classify (e.g., systematic biology, archaeology, paleontology, anthropology), what is collected in the field is turned into science indoors—usually by the same people—in a museum, lab, or herbarium. (Think of these as transhumant sciences.)[3] Other field sciences have split into distinct subdisciplines along the field-lab divide. In ecology, for example, this division is acknowledged in distinct names: "autecology" for the lab discipline, "synecology" for the field. Field scientists deploy laboratory methods and conventions knowing that they thereby make field practices more credible in the world of science. Ecologists are again a good example,[4] or contemporary systematists, when they borrow practices of statistics and molecular genetics. English ethnographers are a more complex case. To justify the dicey practice of embedding lone observers for long periods among isolated exotic peoples (who was to say their reports were true?), they appealed both to a field logic of "heroic" personal virtue and to a laboratory concept of an empathic resident observer as a precise psychological recording instrument.[5] So if practitioners acknowledge and use the categorical distinction of lab and field, why should we not as well? Dynamic tensions between lab and field practices and ideals are constitutive of field sciences. We cannot do without the categories of lab and field: we just need to take care not to take them more rigidly than practitioners themselves do.

Let me turn, then, to the themes that have guided the history of field science since Kuklick and I last reviewed it a little over a decade ago.

Emergent Trends: Vernacular Science

One of these themes is the connection between expert and popular or, less tendentiously, "vernacular" science. Historians of science have long taken an interest in popular science, of course; but with the fading of the epistemic agendas of 1980s science studies, vernacular science has succeeded constructivism as a mainstream interest. It was the leitmotif of the influential 1996 smorgasbord on cultures of natural history edited by Nicolas Jardine, James Secord, and Emma Spary.[6] And the history of all the field sciences, as well as natural history, is benefitting from this turn to the vernacular.

The history of natural history has been especially well developed in the early modern and Enlightenment periods, when it encompassed half the world of natural knowledge, and when no sharp distinction was made between expert and vernacular. Perhaps because it was mainstream science, its historians dealt with it in the mainstream ways of cultural history—as texts, objects, and representations—and not as field science. That has now begun to change, however, most notably in Alix Cooper's book on local "floras" and the various other early modern genres of natural inventory.[7]

Unlike most other historians, Cooper relates literary products to the field practices and local social actors and situations that produced them. The small local floras (plant lists with names and locations but without the customary vernacular lore) were characteristic of the German principalities and were produced, Cooper shows, by local professors of medicine, working alone and on a shoestring, as aids for their students' field excursions (seeing nature for themselves). The mineral lists that appeared later in the century, also in the Germany territories, were similar but arose from the quite different local context of princely and aristocratic interest in mining resources.

Appealing to differences in local contexts, Cooper can also neatly explain why it was that German naturalists ignored the importuning of networkers like Henry Oldenburg to integrate local floras into continental networks: in these locales there was no one to do it, no patrons to support it, and no rewards for doing it. For the same reasons Germans never adopted the English model of county natural-history surveys, which combined botany with geography, archaeology, and local history; nor larger regional surveys, which required royal patronage. And Cooper places this entire story in the context of recurring efforts by continental Europeans to assert the value of "indigenous" nature over the more fashionable exotica of the new worlds being opened up by the maritime states.[8]

Though Cooper's book is not strictly speaking a history of field science, it is a model of how such history could be done: texts, fieldwork, patronage, reward systems, and expert and vernacular culture are all nicely put together. (It is amazing how the old Sci Rev keeps being the nursery of fresh ways of doing history.)

Natural history in the modern period is more difficult to conceptualize. The term has become ambiguous. It is most commonly applied to the residuum of amateur and recreational practices that remained after the breakup of the polyglot empire of natural history into domains of expert science (ornithology, zoology, biogeography, ecology, etc.). Yet all these expert sciences remain connected in vital ways with vernacular traditions—though not necessarily including active participation by amateurs. "Natural history" can thus also refer to the whole complex terrain that once was whole and undivided. An even-handed treatment of expert and vernacular is not a bad thing—so long as the expert sciences do not end up being marginalized in an intellectual field dominated by literary and cultural historians.

One way to deal with modern "natural history" is to focus on the new varieties of social connection that were invented in the nineteenth century to replace active participation and bridge the newly opened divide between expert and vernacular cultures. Consumption of prepackaged science—bird and picture books, nature stories, films, tours, and so on—has become the most common form of vernacular participation in natural history. But consumption need not be passive. Birding, for example, is itself a field practice, though for the purpose of pleasure rather than producing science.[9] A recreational interest in natural history can also intensify into active patronage of field research. Museums (especially in the United States) depended on the patronage of nature lovers for programs of global collecting expeditions, beginning around 1900 with gathering raw material for diorama exhibits. Geographers also had, and still have, a strong connection with popular patronage.[10]

In a few field sciences, direct amateur participation persists (though carefully regulated): for example, bird banding, archaeology, restoration ecology, and fossil hunting.[11] And practical field sciences like horticulture, agricultural extension, or forestry sustain roles for career scientists that are both expert and vernacular—because these are service sciences. Such mixed practices become two-way streets of influence. Katherine Pandora, for example, has shown how the horticulturist and vernacular theorist Luther Burbank was for many Americans the quintessential "scientist" because he did not separate his science from everyday concerns of domestic life and child rearing. Christopher Henke has shown in his study of agricultural field trials how participatory experiments in the field transgress some of the most fundamental principles of laboratory practice in the interest of making the results credible to farmers. (For example, field hands collect "data"—graded produce—and farmers decide when experiments end.)[12]

The relation between expert and vernacular in natural history remains a contentious issue in democratic societies, in which social access to privileged categories is a political issue. Although the "trickle-down" of expert into popular science has been declared not to exist—it does exist, just not universally—the "trickle-up" of vernacular into expert practices is highly approved and a topic of lively current interest. Besides Henke's study of celery-field experiments, Christian Young has shown how the concept of carrying capacity entered ecology from practical range management, not the other way around. Historians have found instances of trickle-up in industrial forestry, river fisheries, and wildlife ecology.[13] It is what we would expect in "applied" or economic sciences generally, because these are more hospitable to mixing of expert and vernacular than, say, academic ecology, which tends to steer clear of popular science.[14] In any case, these are strategic sites to see how in the modern world expert and vernacular cultures have developed mutually (if uneasily) supportive relations.

Emergent Trends: The Long Degree

But perhaps the most noteworthy recent trend in the history of field science is the burgeoning interest in large-scale scientific practices: in sciences of space, science pursued on a regional or global scale. It's history of the "long degree," one might say (with apologies to the old *annaliste* historians of the *longue durée*), and it's becoming a major feature of the history of field science. (If a deep time horizon makes good history, so also does a wide spatial one.)

Of the sciences of the long degree, geology, especially historical and stratigraphic geology, has long attracted historians' attentions. Martin Rudwick's monumental study of historical geology includes a good deal on geology as field science.[15] And field practices are the central feature of Simon Knell's detailed study of English geological and paleontological collecting—by local collectors as well as career field geologists like William Smith and John Phillips. In its treatment of the nitty-gritty of fieldwork and the connection of collecting to the building of provincial museums, Knell's work is exemplary.[16] Recent studies of national geological surveys in Britain by David Oldroyd and Graham McKenna, and in Portugal by Ana Carneiro, also give significant space to the organization and practice of fieldwork.[17]

Likewise with the related science of tectonics, which like stratigraphy involves collecting of data (rather than specimens) on an oceanic and global scale. Once mainly the object of intellectual history, tectonics is increasingly also treated as field practice. Naomi Oreskes's recent history of continental drift shows how individuals' acceptance or rejection of the idea correlates with differences in field practice. Simon Lamb's account of his work on the tectonic history of the Andes affords a vivid insider's view of tectonics as fieldwork.[18]

But if interest in the history of physical geology has not waned in recent years, neither has it grown as vigorously as one might have expected. Nor have economic geology and especially physiography—the science of landforms—proved attractive to historians of field science; despite their importance and inherent intellectual interest, they seem to lack topical appeal.[19]

Rather, history of earth science has developed most vigorously in less earthy and material lines: most notably in geodesy—the science of measuring the Earth's girth and shape. A cluster of recent prize-winning or best-selling books is indicative of the scholarly and popular interest in how the world is measured. These include Mary Terrall's study of Maupertuis, who showed the Earth to be oblate; Ken Alder's, of the meridional measurement that defined the standard meter; and Daniel Kehlmann's novel about Alexander von Humboldt and Carl Friedrich Gauss—the one a traveler and the other a closet theorist, but both avid measurers. Though it is fictionalized "history," Kehlmann's book—a runaway best seller in Germany—captures vividly (and readably) the German obsession with measuring the whole world and all that it contains.[20]

The most striking recent trend, however, is the interest in the sciences—oceanography, meteorology, climatology—that deal globally with the elemental substances and forces that constitute and roil the global biosphere—water, clouds, ice, air, heat, cold, currents, tides. (Is this because the biosphere, unlike the subterranean world, is a realm of everyday experience and concern for us?)

The essays in this volume exemplify this trend: Reidy's on tidology as a "science of space," Rozwadowski's on oceanography, and Fleming's on climatology.

And much more has been done in the last ten years: on weather science by Katharine Anderson, Deborah Coen, and others; on oceanography by Rozwadowski in her earlier work; and on the history of the idea of global warming by Spencer Weart, among others. Though traditionally histories in these areas have emphasized organization and politics, these authors treat their subjects as field sciences, with special attention to the practices of collecting data on a global scale and of managing gigantic databases (on this last topic see the work of Paul Edwards).[21]

Other sciences of the biosphere, like glaciology and paleoclimatology, remain understudied, at least by historians of science. Global mapping from space is another fruitful subject for historians of science on the grand scale. Field scientists and popular science writers have been more alert to the inherent interest of these subjects, and their publishing success should alert us to a boat that we historians of field science do not want to miss.[22]

And if interest in geological survey has somewhat cooled since 1996, there has been a compensating, vigorous growth of interest in large-scale surveys of other and more complex kinds. Take geography. Geographical survey encompasses both physical and cultural aspects of locales: their natural endowments as well as patterns of human occupation and exploitation. Several recent histories have focused on how creators of modern nation-states have used geographical survey to give citizens a sense of personal identification with new nations. Leading studies of this pervasive phenomenon include David Matless's study of regional geographical surveys in Britain and Simon Naylor's of geographical surveys in South America.[23] Historians of science have with a few exceptions not sustained an earlier budding interest in social surveys of towns and urban neighborhoods.[24] But the social and human sciences afford such rich material for historians of field science that interest may spontaneously revive. (We are always most engaged by ourselves.)

Land-use survey is perhaps the most complex survey mode, mapping types of agriculture, forests, soils, and water resources in a multipurpose survey serving all human uses. Two recent examinations of this interesting field science stand out: Alex Checkovich's history of New Deal regional survey in the American Middle West; and Helen Tilley's account of the contemporaneous African Research Survey, which was carried out on a continental scale. Both works treat land-use survey not just as political economy or policy history but also as field science, and they dwell on field practice.[25] Though colonial Africa and the industrial-rural United States were very different places, survey science was in practice much the same in both.

I make a similar point in my study of natural-history surveys—that is, specimen collecting for taxonomic and biogeographic science—by American naturalists in the late nineteenth and early twentieth centuries.[26] These surveys were a complex type of field science: geographically extensive and far-ranging, but also locally intensive and exhaustive. (In contrast, exploration is extensive but not intensive; and "labscape" sciences like ecology are locally intensive but not far-ranging.) Survey collectors aimed at complete species inventories, and so they had to travel widely and easily cross-country, yet at the same time to stay put for extended periods at collecting sites. Such a complex mode of field practice has been attainable historically, I argue, only when conjunctions of environmental, cultural, and scientific conditions provided ready access to areas still wild, social support for systematic expeditioning, and career rewards for systematic collecting and data gathering.

Such a conjunction occurred in the late nineteenth century in the post-frontier American West—a brief period of "inner frontiers" between pioneer settlement and mono-crop industrial agriculture, when large areas were still wild yet easily accessible. How well this model applies to natural history surveys by Europeans in other parts of the world not favored by American collectors (e.g., Siberia and Inner Asia, India, Australia, northern Scandinavia), or to other types of natural inventory, remains to be seen. Since inner frontiers have been a recurring feature of world history in the centuries of sustained population growth and Europe's global expansion, however, the model may have general application. We need more case studies and more comparative work on national traditions of global collecting.

One final thought on sciences of the long degree: these are found not just in the field sciences. Genomics, for example, is representative of a laboratory practice that operates through distributed networks and on a world scale, and generates stupendous accumulations of digital data. Bruno Strasser, who studies DNA data banks, has suggestively likened these digital monsters to natural history collections, and data collecting to collecting specimens in the field.[27] Data banks might also be usefully understood as a virtual long degree—worlds packed into servers. The field sciences may thus afford a strategic vantage for understanding a phenomenon general to all science.

Emergent Trends: Travel and Exploration

Finally there is the subject of travel and exploration—the last of my emergent themes. The role of geology, botany, geography, and natural history in these activities has long been recognized. Exploration for commercial or strategic purposes was often combined with scientific reconnaissance; and recreational travel often involved naturalizing, especially by middle-class travelers for whom it was important that leisure be improving as well as fun.[28] Current enthusiasm for these topics in cultural history, as well as for the history of European imperialism, has pulled historians of field sciences more powerfully in their wake than I would have imagined ten years ago.

Works on science and travel just since 1996 are too numerous to cite fully here, but exemplary studies include those of Michael Bravo, Sverker Sörlin, and others on scientific travel in the far north and arctic; and of Felix Driver, David Arnold, and Graham Burnett on the tropics.[29] There have also been works in a more theoretical vein on the practices of travel and expeditioning: for example, on the existential qualities of spatial separation (Felix Driver), the invention of the expedition as an instrument of fieldwork (Emma Spary), on the construction of "the field" (Alix Cooper), and on the process by which specimens of local value are assimilated into cosmopolitan science.[30]

Recreation and tourism, missionary zeal, and amateur or commercial collecting were no less potent incentives to travel and explore than the hope of profit or power. Field sciences were not just tools of empire but activities that defined a globally inquisitive and restless middle class. An especially rich subject for historical research is the cultural borrowing and translating that has occurred when travel and exploration were pursued for mixed purposes of science, religion, or recreation. In the case of religious travel, Martin Rudwick's think piece on the resemblances of field geology to pilgrimage provides a model.[31] My own work on natural-history collecting shows how field science blended in its practices and meanings with camping, sport hunting, and outdoor recreation. Peder Anker has made a similar case for "science as a vacation" in the history of ecology in Scandinavia.[32]

It is likely that interest in these cultural aspects of field science will strengthen as imperial history moves away from a simplistic model of Western control and expropriation of resources to a more nuanced vision of two-way interactions with indigenous peoples. Whereas earlier studies of imperial science tended to focus on the practical sciences of navigation, geology, and plant prospecting, for example, recent work has inclined more to sciences unconnected with extraction, like natural history or ethnography. Richard Grove's monumental studies of "green imperialism" dwell on the botanical knowledge that European naturalists acquired from indigenes.[33] And as Jorge Cañizares-Esguerra has shown, Spanish scientific expeditions were not just extractive but cultural as well.[34] Two recent studies of the personal relations between European natural-history collectors and their native assistants make the same point on a more intimate, personal scale.[35]

Scientific collecting is an aspect of exploration and travel that has yet to receive the attention it deserves—at least the practice of it. Much has been done on the history of objects and the symbology of collections, far less on collecting as field practice. I have recently reviewed the literature of collecting in paleontology, natural history, archaeology, and ethnography and speculated on how the collecting sciences have been shaped by the inescapable physical necessities of preserving, managing, and ordering large, permanent collections of found objects.[36] Object-based sciences constitute a distinctive family group, inviting comparative study.

There are exemplars of how collecting history can be done, most notably Simon Knell's richly detailed history of fossil collecting and museum building in early nineteenth-century Yorkshire, which deals evenhandedly with collecting and collections. My own study of natural-history survey is likewise an environmental, social, and scientific history of collecting.[37] A comparative study of the collecting practices of Charles Darwin and Alfred Russel Wallace shows how their differences—the one recreational and selective, the other commercial and strenuously intensive—engendered different conceptions of animals, as individuals and as species.[38] A recent collection of essays on the theme of collecting as scientific practice—the first such to appear—augurs an organized interest in collecting history among historians of science.[39]

General historians have also written works that should interest historians of field science. Collecting as an aspect of imperial appropriation of indigenous cultures is a familiar theme in these works. But there is a growing appreciation of the importance of intellectual curiosity in the West's encounter with exotic worlds, as well as of the role of natural history, earth science, and archaeology in forming hybrid colonial cultures.[40] Maya Jasanoff's broad-ranging history of cultural collecting by French and British imperial officials and cultural adventurers gone native—a wonderfully odd and engaging lot—does not deal with scientific collecting. But her treatment is a model for historians of collecting science who want to combine field science and social history.[41] Collecting as field practice is thus beginning to get its due, and it is clearly just the beginning.

Environmental History and Field Science

The themes discussed so far have for the most part "emerged" out of ongoing traditions in the history of science. Others are more deliberate importations from one or another branch of general history. This is an impulse that is strongly felt these days: since history departments are now where historians of science are likely to find employment, we favor subjects that historians will deem worthy and that can be easily spliced into history curricula. Field sciences may benefit from this connection, being less off-putting to historians than, say, chemistry, physiology, or physics.

The organizing themes of the conference from which this volume derives—environmental history and globalism—are both borrowed. They are also, as it happens, the very ones that Jim Secord, in his wrap-up essay for the 1996 volume on cultures of natural history, pointed to as being the most likely to bear fruit. "The history of natural history needs to become part of environmental history," Secord wrote. "And . . . the history of natural sciences needs to take on an international perspective, centered on the global economy of market relations."[42] The organizers of the present volume did not knowingly borrow these themes—because neither they nor I recalled them. But if his words were a bit too prophetic in 1996 to be properly noticed, they are right on the money now as an agenda for the history of the field sciences.

I will have most to say here on connecting with environmental history: because I have done it myself and found it most fruitful, and because at the 2007 conference it was an issue that aroused sustained and earnest discussion. It is a natural alliance. Since field sciences are by definition carried out in some natural or human environment, how can we not make common cause with environmental historians? The only question is, what particular topics will afford the most mutually beneficial symbiosis? I will suggest three—working landscapes, regions, and human ecology—but there are doubtless others no less fruitful.

Working landscapes—environments in which humans are a dominant presence—are likely to be an especially fruitful subject for historians of field science. These have been hugely popular with environmental historians since the mid-1990s, when "wilderness" and wilderness preservation became politically toxic for environmentalists and historians alike (because seemingly dismissive of inhabitants' need to make livings).[43] So environmental historians turned to cities, suburbs, parks, mining and logging areas, and rural agricultural and recreational landscapes. (Landscape historians have always concentrated on man-made, "vernacular" landscapes.) This turn also derived force from Richard White's capacious concept of "knowing nature through work," exemplified in his history of the Columbia River and its varied human users.[44]

Exactly what counts as "working" environment is of course not a matter for a priori definition but something that must emerge from the variability within the category. Since there is really no place on earth that humans have not altered, any place could in principle be considered a "working" landscape. That of course would be self-defeating. The trick is to discover where the category has real analytic leverage and where it shades over into mere narrative trope.

My own preference, which stems from my investment in the history of ecology and natural history, is for environments that are substantially natural yet fairly densely settled: rural landscapes, exurbs, what I have called "labscapes"—places where lab-like practices can be done in nature—as well as "inner frontiers"—where islands of nature survive in the interstices of continental transport webs.[45] But field sciences are carried on in a wide range of working landscapes. Ethnographers, for example, have operated in urban neighborhoods and corporate parks as well as rural hamlets and remote outbacks. Archaeological sites are often also working landscapes, owing to the human preference for continuous or recurring occupation of favored sites. Geologists, paleontologists, glaciologists, oceanographers, and naturalist-collectors venture into some marginally inhabited places; yet even there field scientists depend on networks of communication, transport, and supply.[46]

An analytic of "working landscapes" has many advantages for historians of field science. It impels us to treat field sciences as modes of work, and to think about the ways that science workers engage with the many other kinds of workers who also operate in such environments. It is a category that brings together expert and vernacular ways of knowing. It invites us to think of the conventions of human economy and culture as ecological principles.

The ecological principles of working landscapes are in fact equally natural and cultural, because in such places humans are the dominant species, and culture is our ecology. In field experiments, for example, human actors and human actions are not extraneous elements but essential parts of the experimental situation. In Christopher Henke's case of the agricultural field trial, the measure of ecological production was economic—boxes of graded celery—not biomass, as would be the case in a laboratory trial. Or consider Naomi Oreskes's study of the U.S. Navy's experiments on measuring ocean water temperature by sonic methods. Their experiment quickly ran afoul of green politics (would sonics disrupt whales' ability to navigate?), and oceanographers were faulted for their insensitivity to a legitimate public interest in their experiment. But their failure of imagination was quite understandable: they had simply adopted a concept of "control" from the lab—that singular place where humans must be excluded from experiments. It takes effort to see that in a working seascape experimental "control" embraces human as well as physical and biotic elements. In the field, green politics was not an externality but part of the experiment.[47]

Stuart McCook's essay in this volume on coffee rusts and the ecology of global epidemics affords another such instance. Before coffee growing was an interconnected global industry, the habitats of plants and parasitic rusts were physical and biotic (temperature, humidity, sun, shade). But in the working landscapes of coffee plantations, habitats also included the human elements of trade, assisted migration of cultivars and pathogens, and the peculiar cultural logic of market economics. It's the same with all the plants and animals of working landscapes: their ecology is in part human ecology. Thus in the sciences of working landscapes, humans are not intruders to be excluded as externalities but rather objects of investigation along with other natural things. The rules and customs of science change in working landscapes, which is one reason why the field sciences are so interesting.

A second defining—and eminently poachable—feature of environmental history is the region. It is striking how many of the most successful works of environmental history take regions as their units of study.[48] Regions make such good subjects because they are natural units, both ecologically and culturally, and are an optimal size for substantial but not sprawling case studies. They are a natural unit of historical analysis and narrative for historians who make it their particular business to unite nature and culture—like environmental historians and historians of field science.

Yet with a few exceptions historians of science have not taken advantage of regions as objects of study. The most notable exception is the well-developed history of English provincial science: it focuses on the Midlands, where cultural interest in local geology and natural history was part of the great social and cultural transformations of industrialization. Similar conjunctures doubtless occurred elsewhere, so the English case is not an exception but a model for further work on science regions. Another such model is Michael Smith's book on the field sciences in California from the gold rush to Hetch-Hetchy. In this case it is not industrial transformation but frontier settlement that gave science a distinctive regional quality.[49] It is a highly imitable model that was inexplicably not imitated.

There are, however, new signs of interest in regions currently among historians of science who have the now-abundant achievements of environmental historians to inspire them. Jeremy Vetter's work in progress on the U.S. Great Plains and Rocky Mountain regions is the most deliberately regional and modeled on environmental history. The advantages of a regional unit are evident here. Within a regional frame Vetter can treat all the field sciences, from botany and agriculture to paleontology and geology, without becoming a captive specialist in any one. He can also relate the development of the sciences to the economic and social history of his region in the period of Euro-American settlement, revealing how science was connected to the varied activities of daily life.[50]

The concepts of working landscape and region are alike in making it easy—indeed, unavoidable—to treat nature, ecology, and human culture as a single subject. Working landscapes are by definition places of human ecology; regions are de facto cultural and natural units. It is not as easy as we might think to keep nature and culture together: though they are inseparable in the real world, the customs of academic knowledge making tend to sort them into separate analytic categories. As Sharon Kingsland has observed (and deplored), ecologists have typically avoided humanized environments (no doubt because by the criteria of lab science these are tainted by human presence, and ecologists are judged by lab standards).[51] Likewise, ecological explanations make historians nervous, doubtless because these seem "deterministic" and beyond the ken and control of historical expertise. (Cultural determinism we humanists do not fear; natural "determinisms" we fear greatly.) The customs of academic attack and defense thus favor "pure" tribal methodologies over those with ambiguous, dual standards. Kingsland appeals for a history of ecology that includes human ecology. I would extend this appeal to the history of all field sciences.

Globalizing Field Science

Global history—the second theme of this volume—is, like the first, a deliberate import from a more powerful sister discipline. Historians are going pell-mell for globalism, as is the world. History confined to national boundaries is out, we now believe, and the future lies not in the West but the South and East. For historians a global reach is de rigueur, and historians of science likewise hope to sail this prevailing wind to a historiographic new world.

Though the rewards of global history may prove to be generally harder to realize than its promoters think, there are particular subjects in the history of science that lend themselves to a global approach. One of these is the pervasive phenomenon of global networks—of trade, travel, exploration, knowledge, and power. In these distributed networks, botany, geology, and other field sciences have played a significant role. Richard Grove, for example, has shown how networks of trade and empire gave Europeans access to indigenes' ethnoscientific knowledge, and gave them clear evidence of the disastrous ecological consequences of intensive human land use.[52] Historians of the field sciences have also begun to make signal contributions to this subject: Emmy Spary, on the Jardin du Roi; Fa-ti Fan, on the networks of British naturalists who followed the China trade to the sources of supply of specimens in Chinese entrepôts (being unwelcome aliens, they were banned from collecting in the field).[53] Lisbet Koerner has shown how the Linnaean network of botanical collectors was embedded in a mercantile system of horticultural prospecting and acclimatization. Staffan Müller-Wille traces a local plant through a Linnaean network to show that its assimilation into cosmopolitan science occurred not just at a "center of calculation" (à la Bruno Latour) but at every point in its journey.[54]

Global sciences also exemplify a spatially distributed form of scientific production that, unlike the tight communities of disciplines, relies on the "strength of weak ties."[55] Meteorological and geophysical sciences have long been organized in this distributed way, and genome projects are a more novel type on the laboratory side. The field sciences, with their deep traditions of global networking, are strategic sites for understanding a form of organization that is widespread and spreading.

Another advantage of the "global" rubric is that, in contrast to the "imperial" rubric, it is confined neither to the era of European empires nor to simplistic, one-way models of imperial metropole and exploited peripheries. (There were hints in the 2007 conference of a certain exasperation with the mantra of imperial dominion. "I'm sick of doing only empire," a senior historian expostulated at one point, to murmurs of surprise and apparent approval.) Globalism affords a more agreeably polycentric vision of human interactions.[56]

It seems certain that history of science will continue to go global, and that history of field science can only benefit—if only because its subjects are themselves globally distributed. I've discussed above the sciences of the long degree, which deal with air, winds, water, and tides. But molecules also go global, some notable examples being carbon dioxide, industrial effluvia, and radioactive atoms. And there are global biological phenomena: faunal migrations of birds, fishes, and whales; and opportunistic invasive species, especially our own "portmanteau biota" of commensal species that travel with us and assist in demographic takeover. And there is acclimatization (the "paradigmatic colonial science," Michael Osborne has called it); it is another science of global migrations.[57] One may hope as well that globality will breathe new life into the history of biogeography, a subject of great interest that after foundational work by Janet Browne and Jane Camerini has unaccountably languished.[58]

And most especially, there is us—*Homo sapiens*, the most restless, footloose, uncontainable species of all. Lynn Nyhart's essay in this volume on the scientists who study human migrations, and on the ways in which the experience of migration can affect scientific thought, opens a window on a large and fruitful subject. If we can make travel into a historical cottage industry, why not also the peregrinations of scientists who have studied human restlessness?[59] Our propensity to go global is after all a defining element of our own, human, ecology.

It is no accident, I think, that the themes that have been most fruitfully borrowed from environmental and global history—working landscapes, regions, networks, sciences of the long degree—are all spatial categories. Place has long been a productive analytic in science studies, though so far primarily in local case studies. But that limitation was a local contingency of our own disciplinary development, and scaling up presents no inherent difficulty. And larger conceptions of place will aid us in breaking through (or reinvigorating) categories of discipline and nation in which we think (perhaps prematurely) there is nothing new to find.

Themes of Our Own

Historians of field science also have rich historiographic resources closer to home, in the traditions of big-picture history of science, and science studies. We need not poach or borrow, just read widely in our own discipline.

The most enduring of our big-picture themes is the nature of the great scientific epochs of Scientific Revolution, Enlightenment, and modernity (and perhaps postmodernity); and of the transitions from one era to the next. Whether these categories are real and useful is debated (I think they are both); what is not disputable is that the field sciences have had little part in this *longue durée* history. That is changing, however, most notably among historians of the early modern period, who are coming to believe that natural history played a far more significant role in the Scientific Revolution than has hitherto been appreciated.

Harold Cook has been the most persistent and forceful in propounding this thesis, for over a decade in essays and, in great detail, in his recent book on Dutch trade and natural history. Because natural history dealt with material objects rather than the deductive systems of natural philosophy, it was an area of intellectual life where a naturalistic, rather than a religious or symbolic, interest in nature could take root—and did.[60] A similar argument has been made by Antonio Barrera-Osorio in his work on the early Spanish empire in the Americas. Empirical science, he boldly asserts, "resulted from the commercial and imperial scientific activities of the sixteenth and seventeenth centuries. The early Scientific—that is, empirical—Revolution that took place in Spain became a key element in later developments [in Britain and Italy]."[61] In other words, naturalistic and empirical habits of mind that predate the more familiar revolution in natural philosophy derived less from philosophy than from European commerce and empire building.

This view resembles the long-familiar idea that exotic plants and animals from the new worlds broke open the fixed categories of Renaissance natural history. But Cook and Barrera-Osorio go far deeper, into the social history of European encounters with exotic natural and human worlds. Cook, for example, highlights the intensely naturalistic and empirical culture of Dutch traders. They had powerful financial reasons to have accurate, detailed, and realistic knowledge of natural stuffs and objects—things of commerce. Financial distress awaited those whose empirical knowledge was imperfect. Barrera-Osorio similarly points to the powerful practical incentives for Spanish imperial officials to have a realistic empirical knowledge of their New World domains. (Natural philosophers, in contrast, had no such incentives, so could dwell happily in abstract worlds of mind.) As a result, natural history in the broad sense became the leading edge of a wholesale revolution in Europeans' approach to the natural world. (Exactly how empiricist, thing-y cultures of trade and imperial administration reshaped natural philosophy remains to be worked out.)

I find these arguments persuasive and suggestive of even wider revisions in big-picture history. If natural history was so central to the Scientific Revolution yet so invisible to generations of historians, might it have played a comparable role in the other great transitions—from early modern to Enlightenment, and from Enlightenment to modernity—that we have simply failed to make visible? Is the elephant in these rooms as well? There are reasons to think it might be so.

It seems clear, for example, that natural history was especially well suited to the mode of organizing science favored by Enlightenment savants: namely, a union of expert and vernacular elements in institutions in the new "public sphere" (academies, societies, salons, coffee houses).[62] Sociability was the organizing principle of Enlightenment science, and the field sciences—natural history, archaeology, ethnography, paleontology, geography—were subjects around which diverse sociable groups could readily form, because they were accessible and they connected expert learning with lived experience. The rich history of local natural-history societies—a quintessential institution of Enlightenment science—affords evidence for this view.[63]

There are likewise hints that the field sciences played a role in creating typically modern social relations. Consider, for example, the close connection between science and the central institution of modernity—the nation-state. Creators of modern nation-states had to refashion peasants into citizens whose identity was national rather than local or regional; and natural history, geography, and archaeology (as well as language and folkways) were areas of culture in which proto-citizens could identify with both home locales and cosmopolitan states. The German *Heimat* movement is an especially rich case, and other European countries had something like it.[64] Though *Heimat* historians have paid more attention to culture and remembrances than to science, the opportunity for historians of field science is clear enough, and some have begun to take advantage of it, most notably Lynn Nyhart in her recent book on German biology.[65]

Science in the modern period is also defined by its intensive use of social organization and resources, and its practices of systematic fact gathering and data management. This was the empiricism of preceding periods enormously amplified and empowered by national organization and state funding. Here too a case can be made that it was the field sciences that led the way. It was the collecting sciences of geological and natural-history survey, and taxonomy—not physiology or physics—that exemplified the data-driven, money- and organization-intensive science of the Age of Progress.[66] The field sciences were the "big science" of their day. Museums were initially the premier scientific institutions in modernizing universities; only gradually did labs become a separate and then a dominating architectural and intellectual feature of university quarters. Field sciences like geography and natural history were crucial as well in creating institutions of popularized science that connected producers, who in the modern way wrote only for each other, and the reading, voting publics on whose support expert science now depended.

These thoughts are bare-faced speculations, admittedly. But if they lie anywhere near the truth, then historians of the field sciences have in their possession intellectual levers that could turn big-picture historiography of science on its head. The possibility seems worth pursuing.

The sociological or science-studies side of our field also offers opportunities for historians of field sciences in the issues of "construction" and practice. We have amassed a considerable knowledge of scientific practices in various venues—lab, garden, field, museum, study. Yet there has been little effort to treat the diversity of practices systematically and comparatively, as a taxonomist or biogeographer would treat natural diversity. How many kinds, or species, of practice are there? How were these constructed as lab or field sciences, and how have they evolved? Where is our Linnaeus or, better still, our Darwin of the history of practice?

It was in the nineteenth century that the field sciences were "constructed" in their modern form—that is, transformed from more-or-less indoor pursuits to sciences actively pursued in the field. More precisely, practitioners who had relied for their evidence on a lesser caste of travelers and collectors were succeeded by those who did their own collecting and observing in the field. Philosophical debate on the merits and shortcomings of cabinet and field practice was lively and ongoing. But the shift from armchair to field was probably driven by ever-rising standards of exact empirical evidence in science generally. (Scientists were better gatherers and observers because they knew what to gather and observe.) Anthropology is the best-known case; but geologists and paleontologists, systematic biologists, geographers, biogeographers, and archeologists experienced a similar transition, at different times and in different ways.[67] Comparative study would repay the effort.

Though none of the essays in this book treats the diversity of field practice as a general issue, many display a nascent interest in defining basic types. Such attempts bubbled up spontaneously in papers and discussion, as somewhat playful forays in inventing names for distinctive modes of field practice. The idea of "extreme science" arose at several points: that is, science carried out in difficult or life-threatening circumstances such as high altitudes (Vetter) or ocean depths (Rozwadowski), where the smallest misstep can be fatal. The term "extreme" has sporting connotations (bungee jumping, skiing off cliffs), but personal risk taking has epistemic meaning as well. As Bruce Hevly has argued in the case of the "heroic science" of alpine glaciology, researchers who put their lives on the line to do science are taken to be unimpeachable scientific witnesses.[68] The same logic of credibility operates in ethnography and doubtless many other field sciences.[69]

"Spatial science" is another such category, the term coined by Michael Reidy to designate a family of practices carried out over wide spaces—sciences of the long degree. And Emily Brock devised the category of "restorative" field practices for a mode of industrial forest management that takes ecological restoration as an end along with efficient production of board feet.[70] The term would also apply to the practices of restoration ecology and landscape ecology, and to any practice that combines short-term economic with long-term ecological purposes. In a world of working landscapes, there are likely to be many kinds of "restorative" science, and having the pigeonhole will make it easier to see them.

"Citizen science" was yet another term that came up in the discussion of several papers, to designate practices in which lay persons participate along with expert practitioners: Native Americans in polar research, for example (Michael Bravo); and farmer archaeologists (J. Conor Burns). Jeremy Vetter's rancher paleontologists are yet another instance.[71] As environmental scientists learn to engage active public participation, "citizen science" will doubtless evolve many variants. I have already referred to my own designation of "collecting sciences" as a distinct family of object-centered sciences exemplified by systematic biology, archaeology, anthropology, and paleontology.

Another such rubric, also my invention, is "residential science." By this I mean a mode of expert practice that requires knowledge of some particular locale as intimate and complete as the knowledge of long-term residents. I first used the term in my study of natural history survey, to differentiate the work of visiting survey scientists from that of local residents. The one had cosmopolitan knowledge of taxonomy and biogeography; the other, knowledge of local particulars—where species animals and plants lived and could be found and collected.[72] Visiting scientists could in time acquire residential knowledge, but found it more efficient to rely on actual residents. It subsequently dawned on me, however, that this division of labor was just one variant of "residential," and that there were modes of fieldwork that combined expertise and "residential" know-how in a single scientific role.

Wildlife ecology is a residential science in this broader sense. In his landmark studies of predation on quail and muskrat, for example, the ecologist Paul Errington kept track of every individual animal in a study area over a period of years, charting population ups and downs and ascertaining the fate of each animal (death by predation, starvation, poaching). Such work required both cosmopolitan learned knowledge of animal biology and the intimate, local know-how that could only be acquired by becoming as much a resident of a place as the animals themselves. Practitioners of "residential" science in effect lived and worked inside the objects of their study.[73] Errington did not literally reside in his study areas, but others did: for example, social anthropologists, who lived for extended periods among "their" tribes; or homebound wives like Althea Sherman or Margaret Nice, whose ethological studies of chimney swifts and song sparrows were literally "backyard" science. Konrad Lorenz did his famous research on domesticated birds while residing at his family estate in the Austrian countryside.[74] Still other variants are Jane Goodall's intimate, long-term study of the Gombe chimps, Peter and Rosemary Grant's microevolutionary study of Galapagos finches, and settlement-house urban social research.[75] It is a large and diverse family of field sciences, which all deal with the intricacies of social behavior and relations in animal (including human animal) communities in their environmental contexts, and which are best studied by resident observers.

We don't yet know just how many definite types of scientific practice there are; it will take systematic study to find out. It is clear, however, that it is in the field sciences that we will find the greatest diversity of practices. This variability presumably arises because the field sciences are carried out in varied natural situations by socially mixed groups of practitioners (rather than in radically simplified, socially homogenized, and well-regulated places like labs). Varied conditions of work invite and reward variations in practice; and field practitioners are less likely to be penalized for deviance from a standard set of rules—because in the field there is no one standard. The field is thus a nursery of scientific diversity, as some places in nature (isolated lakes, islands) are nurseries of new species. If it is a natural history of practices that we are after, the field is the place to do it.

In sum, the history of field sciences appeals not just because these sciences are fresh subject specialties, but because they afford strategic research sites for developing ideas that apply generally to the history of science and that all historians of science will want to know about. The field sciences are no longer a peripheral terra incognita, but the inner frontiers of a fully cosmopolitan history of science.

Notes

1. Henrika Kuklick and Robert E. Kohler, eds., *Science in the Field, Osiris* 11 (1996).
2. Thomas F. Gieryn, *Cultural Boundaries of Science: Credibility on the Line* (Chicago: University of Chicago Press, 1999); and Robert E. Kohler, *Landscapes and*

Labscapes: Exploring the Lab-Field Border in Biology (Chicago: University of Chicago Press, 2002).

3. Robert E. Kohler, "Finders, Keepers: Collecting Sciences and Collecting Practice," *History of Science* 45 (2007): 1–27.
4. Kohler, *Landscapes and Labscapes*. Lab scientists likewise knowingly borrowed elements of field practice. See, for example, Sharon E. Kingsland, "Frits Went's Atomic Age Greenhouse: The Changing Landscape of the Lab-Field Border," *Journal of the History of Biology* 42 (2009): 289–324; and Robert E. Kohler, "Labscapes: Naturalizing the Lab," *History of Science* 40 (2002): 473–501.
5. Henrika Kuklick, "Personal Equations: Reflections on the History of Fieldwork, with Special Reference to Sociocultural Anthropology", *Isis* (in press); Kuklick, "After Ishmael: The Fieldwork Tradition and Its Future," in *Anthropological Locations: Boundaries and Grounds of a Field Science*, ed. Akhil Gupta and James Ferguson (Berkeley and Los Angeles: University of California Press, 1997), 47–65; and Kuklick, "Fieldworkers and Physiologists," in *Cambridge and the Torres Strait*, ed. Anita Herle and Sandra Rouse (Cambridge: Cambridge University Press, 1998), 158–180. My thanks to the author for showing me her unpublished essay.
6. Nicholas Jardine, James A. Secord, and Emma C. Spary, eds., *Cultures of Natural History* (Cambridge: Cambridge University Press, 1996).
7. Alix Cooper, *Inventing the Indigenous: Local Knowledge and Natural History in Early Modern Europe* (Cambridge: Cambridge University Press, 2007); also Cooper, "'The Possibilities of the Land': The Inventory of 'Natural Riches' in the Early Modern German Territories," in *Oeconomies in the Age of Newton*, ed. Margaret Schabas and Neil De Marchi (Durham: Duke University Press, 2003), 129–153.
8. On the newly opened gap between expert science and public, see Steven Shapin, "Science and the Public," in *Companion to the History of Modern Science*, ed. Robert C. Olby and others (London: Routledge, 1990), 990–1007.
9. The model is David E. Allen, *The Naturalist in Britain: A Social History* (1976; repr., Princeton University Press, 1994). Exemplary recent works include Mark V. Barrow Jr., *A Passion for Birds: American Ornithology After Audubon* (Princeton: Princeton University Press, 1998); and Helen Macdonald, "'What Makes You a Scientist Is the Way You Look at Things': Ornithology and the Observer, 1930–1955," *Studies in History and Philosophy of the Biological and Biomedical Sciences* 33 (2002): 53–77.
10. Robert E. Kohler, *All Creatures: Naturalists, Collectors, and Biodiversity, 1850–1950* (Princeton: Princeton University Press, 2006), chap. 3; and Susan Schulten, *The Geographical Imagination in America, 1880–1950* (Chicago: University of Chicago Press, 2001).
11. For example, Barrow, *Passion for Birds*; Barrow, this volume; Jeremy Vetter, this volume; and Vetter, "Cowboys, Scientists, and Fossils: The Field Site and Local Collaboration in the American West," *Isis* 99 (2008): 273–303.
12. Katherine Pandora, "Knowledge Held in Common: Tales of Luther Burbank and Science in the American Vernacular," *Isis* 92 (2001): 484–516; and Christopher R. Henke, "Making a Place for Science: The Field Trial," *Social Studies of Science* 30 (2000): 483–511.

13. Christian C. Young, "Defining the Range: The Development of Carrying Capacity in Management Practice," *Journal of the History of Biology* 31 (1998): 61–83; Emily K. Brock, "The Challenge of Reforestation: Ecological Experiments in the Douglas Fir Forest, 1920–1940," *Environmental History* 9 (2004): 57–79; Daniel W. Schneider, "Local Knowledge, Environmental Politics, and the Founding of Ecology in the United States: Stephen Forbes and 'The Lake as a Microcosm' (1887)," *Isis* 91 (2000): 681–705; Albert Way, "Burned to be Wild: Herbert Stoddard and the Roots of Ecological Conservation in the Southern Longleaf Pine Forest," *Environmental History* 11 (2006): 500–526; and Robert E. Kohler, "Paul Errington, Aldo Leopold, and Wildlife Ecology: Doing Science, Living Lives" (unpublished MS, 2010). See also Joshua Blu Buhs, *The Fire Ant Wars: Nature, Science, and Public Policy in Twentieth-Century America* (Chicago: University of Chicago Press, 2004).
14. Perhaps academic ecologists, being more sensitive to their standing among the life-science disciplines (owing to the grinding daily competition for space and funding), are therefore quicker to avoid practices could be construed as less than "pure." See Kohler, *Landscapes and Labscapes*, chap. 6.
15. Martin Rudwick, *Bursting the Limits of Time: The Reconstruction of Geohistory in the Age of Revolution* (Chicago: University of Chicago Press, 2005); and Rudwick, *Worlds Before Adam: The Reconstruction of Geohistory in the Age of Reform* (Chicago: University of Chicago Press, 2008).
16. Simon Knell, *The Culture of English Geology, 1815–1851* (Aldershot, UK: Ashgate, 2000). See also, in a more popular vein, Simon Winchester, *The Map That Changed the World: William Smith and the Birth of Modern Geology* (New York: HarperCollins, 1999).
17. David Oldroyd and Graham McKenna, "Conditions of Employment and Work Practices in the Early Years of the Geological Survey of Great Britain," *Earth Sciences History* 24 (2005): 197–223; and Ana Carneiro, "Outside Government Science, 'Not a Single Tiny Bone to Cheer Us Up!' The Geological Survey of Portugal (1857–1908), the Involvement of Common Men, and the Reaction of Civil Society to Geological Research," *Annals of Science* 62 (2005): 141–204.
18. Naomi Oreskes, *The Rejection of Continental Drift: Theory and Method in American Earth Science* (New York: Oxford University Press, 1999); Oreskes, "Gravity Surveys in the 'Permanent' Ocean Basins: An Instrumental Chink in a Theoretical Suit of Armor," in *Oceanographic History: The Pacific and Beyond*, ed. Keith R. Benson and Philip F. Rehbock (Seattle: University of Washington Press, 2002), 502–510; and Simon Lamb, *Devil in the Mountain: A Search for the Origin of the Andes* (Princeton: Princeton University Press, 2004).
19. Paul Lucier, "A Plea for Applied Geology," *History of Science* 37 (1999): 283–318; and Lucier, *Scientists and Swindlers: Consulting on Coal and Oil in America, 1820–1890* (Baltimore: Johns Hopkins University Press, 2008).
20. Mary Terrall, *The Man Who Flattened the Earth: Maupertuis and the Sciences of the Enlightenment* (Chicago: University of Chicago Press, 2002); Ken Alder, *The Measure of All Things: The Seven-Year Odyssey and Hidden Error that Transformed the World* (New York: Free Press, 2002); and Daniel Kehlmann, *Measuring the World*, trans. Carol Brown Janeway (New York: Pantheon, 2006).
21. On weather science: Katharine Anderson, *Predicting the Weather: Victorians and the Science of Meteorology* (Chicago: University of Chicago Press, 2005); James R.

Fleming, Vladimir Jankovic, and Deborah R. Coen, eds., *Intimate Universality: Local and Global Themes in the History of Weather and Climate* (Sagamore Beach, Mass.: Science History Publications, 2006); Deborah R. Coen, "Scaling Down: The 'Austrian' Climate Between Empire and Republic," in Fleming, Jankovic, and Coen, *Intimate Universality*, 115–140; Simon Naylor, "Nationalizing Provincial Weather: Meteorology in Nineteenth-Century Cornwall," *British Journal for the History of Science* 39 (2006): 407–433; Jan Golinski, "American Climate and the Civilization of Nature," in *Science and Empire in the Atlantic World*, ed. James Delbourgo and Nicholas Dew (New York: Routledge, 2008), 153–174; Paul N. Edwards, "Representing the Global Atmosphere: Computer Models, Data, and Knowledge About Climate Change," in *Changing the Atmosphere: Expert Knowledge and Environmental Governance*, ed. Clark A. Miller and Paul N. Edwards (Cambridge: MIT Press, 2001), 31–65; and Mark Monmonier, *Air Apparent: How Meteorologists Learned to Map, Predict, and Dramatize the Weather* (Chicago: University of Chicago Press, 1999). On oceanography: Helen M. Rozwadowski, *Fathoming the Ocean: The Discovery and Exploration of the Deep Sea* (Cambridge: Harvard University Press, 2005); Michael S. Reidy, *Tides of History: Ocean Science and Her Majesty's Navy* (Chicago: University of Chicago Press, 2008); and Ronald Rainger, "Oceanography and Fieldwork: Geopolitics and Research at the Scripps Institution," in *Museums and Other Institutions of Natural History: Past, Present, and Future*, ed. Alan E. Leviton and Michele L. Aldrich (San Francisco: California Academy of Sciences, 2004), 185–208. On global warming: Spencer R. Weart, *The Discovery of Global Warming* (Cambridge: Harvard University Press, 2003).

22. Peter Knight, "Glaciers: Art and History, Science and Uncertainty," *Interdisciplinary Science Review* 29 (2004): 385–393; Diarmid A. Finnegan, "The Work of Ice: Glacial Theory and Scientific Culture in Early Victorian Edinburgh," *British Journal for the History of Science* 37 (2004): 29–52; Richard B. Alley, *The Two-Mile Ice Machine: Ice Cores, Abrupt Climate Change, and Our Future* (Princeton: Princeton University Press, 2000); and Mark Bowen, *Thin Ice: Unlocking the Secrets of Climate in the World's Highest Mountains* (New York: Henry Holt, 2005).
23. David Matless, "Forms of Knowledge and Forms of Belonging: Regional Survey and Geographical Citizenship," in *The City After Patrick Geddes*, ed. Volker Welter and James Lawson (Oxford: Peter Lang, 2000), 91–112; and Simon Naylor and G. A. Jones, "Writing Orderly Geographies of Distant Places: The Regional Survey and Latin America," *Ecumene* 4 (1997): 273–299.
24. Mark Freeman, "The Provincial Social Survey in Edwardian Britain," *Historical Research* 75 (2002): 73–90.
25. Alex Checkovich, "Mapping the American Way: Geographical Knowledge and the Development of the United States, 1890–1950" (PhD diss., University of Pennsylvania, 2004); and Helen Tilley, "African Environments and Environmental Sciences: The African Research Survey, Ecological Paradigms, and British Colonial Development, 1930–1940," in *Social History of African Environments*, ed. William Beinart and JoAnn McGregor (Oxford: James Currey, 2003), 109–130.
26. Kohler, *All Creatures*. Though based in U.S. institutions, these surveys operated worldwide.
27. Bruno Strasser, "Collecting and Experimenting: The Moral Economies of Biological Research, 1960s–1980s," in *History and Epistemology of Molecular Biology and*

Beyond: Problems and Perspectives, ed. Soraya de Chadarevian and Hans-Jörg Rheinberger (Berlin: Max Planck Institute for the History of Science, 2006), 105–123.

28. Cindy Aron, *Working at Play: A History of Vacations in the United States* (New York: Oxford University Press, 1999); and Peter Bailey, *Leisure and Class in Victorian England: Rational Recreation and the Contest for Control, 1830–1885* (London: Routledge and Kegan Paul, 1978).
29. Michael Bravo and Sverker Sörlin, eds., *Narrating the Arctic: A Cultural History of Nordic Scientific Practices* (Canton, Mass.: Science History Publications, 2002); Sörlin, "Ordering the World for Europe: Science as Intelligence and Information as Seen from the Northern Periphery," *Osiris* 15 (2001): 51–69; Per Eliasson, "Swedish Natural History Travel in the Northern Space: From Lapland to the Arctic, 1800–1840," in Bravo and Sörlin, *Narrating the Arctic*, 125–154; Felix Driver and Luciana Martins, eds., *Tropical Visions in an Age of Empire* (Chicago: University of Chicago Press, 2005); Driver, *Geography Militant: Cultures of Exploration and Empire* (Oxford: Blackwell, 2001); David Arnold, *Tropics and the Traveling Gaze: India, Landscape, and Science* (Seattle: University of Washington Press, 2006); and D. Graham Burnett, *Masters of All They Surveyed: Exploration, Geography, and a British El Dorado* (Chicago: University of Chicago Press, 2000).
30. For example, Felix Driver, "Distance and Disturbance: Travel, Exploration, and Knowledge in the Nineteenth Century," *Transactions of the Royal Historical Society* 14 (2004): 73–92; James Clifford, "Spatial Practices: Fieldwork, Travel, and the Disciplining of Anthropology," in *Anthropological Locations: Boundaries and Grounds of a Field Science* (Berkeley and Los Angeles: University of California Press, 1997), 185–222; Emma Spary, "L'invention de l'expédition scientifique: L'histoire naturelle, Bonaparte, et l'Egypte," in *L'Invention Scientifique de la Méditerranée*, ed. Marie-Nöelle Bourguet and others (Paris: Ecole des Haute Études en Science Sociales, 1998), 19–38; Mary Terrall, "Heroic Narratives of Quest and Discovery," *Configurations* 6 (1998): 223–242; Alix Cooper, "From the Alps to Egypt (and Back Again): Dolomieu, Scientific Voyaging, and the Construction of the Field in Eighteenth-Century Natural History," in *Making Space for Science*, ed. Crosbie Smith and Jon Agar (New York: St. Martin's Press, 1998), 39–63; Marie-Nöelle Bourguet, Christian Licoppe, and H. Otto Sibum, eds., *Instruments, Travel, and Science: Itineraries of Precision from the Seventeenth to the Twentieth Century* (London: Routledge, 2002); and Staffan Müller-Wille, "Joining Lapland and the Topinambes in Flourishing Holland: Center and Periphery in Linnaean Botany," *Science in Context* 16 (2003): 461–488.
31. Martin Rudwick, "Geological Travel and Theoretical Innovation: The Role of 'Liminal' Experience," *Social Studies of Science* 26 (1996): 143–159. See also Steven J. Harris, "Mapping Jesuit Science: The Role of Travel in the Geography of Knowledge," in *The Jesuits: Cultures, Sciences, and the Arts, 1540–1773*, ed. John W. O'Malley and others (Toronto: University of Toronto Press, 1999), 212–240; and Sujit Sivasundaram, *Nature and the Godly Empire: Science and Evangelical Mission in the Pacific, 1795–1850* (New York: Cambridge University Press, 2005).
32. Kohler, *All Creatures*, chaps. 2–3; and Peder Anker, "Science as a Vacation: A History of Ecology in Norway," *History of Science* 45 (2007): 455–479.

33. Richard H. Grove, *Green Imperialism: Colonial Expansion, Tropical Island Edens, and the Origin of Environmentalism, 1600–1860* (New York: Cambridge University Press, 1995); and Grove, "Indigenous Knowledge and the Significance of South-West India for Portuguese and Dutch Constructions of Nature," in *Nature and the Orient: The Environmental History of South and Southeast Asia*, ed. Richard H. Grove, Vinita Damodaran, and Satpal Sangwan (Delhi: Oxford University Press, 1998), 187–209. See also Londa Schiebinger, *Plants and Empire: Colonial Bioprospecting in the Atlantic World* (Cambridge: Harvard University Press, 2004). Schiebinger focuses on the quest for herbal abortifacients.
34. Jorge Cañizares-Esguerra, "How Derivative Was Humboldt? Microcosmic Nature Narratives in Early Modern Spanish American and the (Other) Origins of Humboldt's Ecological Sensibilities," in *Colonial Botany: Science, Commerce, and Politics in the Early Modern World*, ed. Londa Schiebinger and Claudia Swan (Philadelphia: University of Pennsylvania Press, 2005), 148–165.
35. Nancy J. Jacobs, "The Intimate Politics of Ornithology in Colonial Africa," *Comparative Studies in Society and History* 48 (2006): 564–603; and Patrick Harries, "Field Sciences in Scientific Fields: Ethnology, Botany, and the Early Ethnographic Monograph in the Work of H.-A. Junod," in *Science and Society in Southern Africa*, ed. Saul Dubow (Manchester: Manchester University Press, 2000), 11–41.
36. Robert E. Kohler, "Finders, Keepers."
37. Knell, *Culture of English Geology*. See also Samuel J. M. M. Alberti, "Owning and Collecting Natural Objects in Nineteenth-Century Britain," in *From Private to Public: Natural Collections and Museums*, ed. Marco Beretta (Sagamore Beach, Mass.: Science History Publications, 2005), 141–154; Alberti, "Placing Nature: Natural History Collections and Their Owners in Nineteenth-Century Provincial England," *British Journal for the History of Science* 35 (2002): 291–311; Richard W. Burkhardt Jr., "Naturalists' Practices and Nature's Empire: Paris and the Platypus, 1815–1833," *Pacific Science* 55 (2001): 327–341; Anke te Heesen, "From Natural Historical Investment to State Service: Collectors and Collections of the Berlin Society of Friends of Nature Research, c. 1800," *History of Science* 42 (2004): 113–131; and Jenny Beckman, "Nature's Palace: Constructing the Swedish Museum of Natural History," *British Journal for the History of Science* 42 (2004): 85–111. Works on cabinets of curiosities are too numerous even to begin to cite.
38. Melinda B. Fagan, "Wallace, Darwin, and the Practice of Natural History," *Journal of History of Biology* 40 (2007): 601–635. See also Jeremy Vetter, "Wallace's *Other* Line: Human Biogeography and Field Practice in the Eastern Colonial Tropics," *Journal of History of Biology* 39 (2006): 89–123.
39. Anke te Heesen and Emma C. Spary, eds., *Sammeln als Wissen: Das Sammeln und seine wissenschaftsgeschichtliche Bedeutung* (Göttingen: Wallenstein Verlag, 2001).
40. For example, Chris Gosden and Chantal Knowles, *Collecting Colonialism: Material Culture and Colonial Change* (Oxford: Berg, 2001); Tom Griffiths, *Hunters and Collectors: The Antiquarian Imagination in Australia* (Cambridge: Cambridge University Press, 1996); H. Glenn Penny and Matti Bunzl, eds., *Worldly Provincialism: German Anthropology in the Age of Empire* (Ann Arbor: University of Michigan Press, 2003); and Saul Dubow, "Earth History, Natural History, and Prehistory at the Cape, 1860–1875," *Comparative Studies in Society and History* 46 (2004): 107–133.

41. Maya Jasanoff, *Edge of Empire: Lives, Culture, and Conquest in the East, 1650–1850* (Cambridge: Harvard University Press, 2005).
42. James A. Secord, "The Crisis of Nature," in Jardine, Secord, and Spary, *Cultures of Natural History*, 447–459, on p. 458.
43. Richard White, "From Wilderness to Hybrid Landscapes: The Cultural Turn in Environmental History," *Historian* 66 (2004): 557–564; William Cronon, "The Trouble with Wilderness; or, Getting Back to the Wrong Nature," in *Uncommon Ground: Toward Reinventing Nature*, ed. Cronon (New York: Norton, 1995), 69–90; and White, "'Are You an Environmentalist or Do You Work for a Living?': Work and Nature," in Cronon, *Uncommon Ground*, 171–185.
44. Richard White, *The Organic Machine* (New York: Hill and Wang, 1995).
45. On "inner frontiers," see Kohler, *All Creatures*, chap. 1; on "labscapes," see Kohler, *Landscapes and Labscapes.*
46. For example, Jane Camerini, "Wallace in the Field," *Osiris* 11 (1996): 44–65; Alex Soojung-Kim Pang, *Empire and the Sun: Victorian Solar Eclipse Expeditions* (Stanford: Stanford University Press, 2002); and Kohler, *All Creatures*, chap. 4.
47. Naomi Oreskes, "The Controlled Experiment That Wasn't: Acoustic Tomography of Ocean Climate," lecture at the American Philosophical Society, May 2007. This lecture will be part of a book in progress tentatively titled *Science on a Mission: American Oceanography in the Cold War and Beyond.*
48. An overview of the regional approach can be found in Daniel L. Flores, "Place: An Argument for Bioregional History," *Environmental History Review* 18, no. 4 (1994): 1–18.
49. Michael Smith, *Pacific Visions: California Scientists and the Environment* (New Haven: Yale University Press, 1987).
50. Jeremy Vetter, "The Regional Development of Science: Knowledge, Environment, and Fieldwork in the Central Plains and Rocky Mountains, 1860–1920" (PhD diss., University of Pennsylvania, 2005); and Vetter, "Science Along the Railroad: Expanding Fieldwork in the U.S. Central West," *Annals of Science* 61 (2004): 187–211. Other works with a regional focus include Daniel Goldstein, "Midwestern Naturalists: Academies of Science in the Mississippi Valley, 1850–1900" (PhD diss., Yale University, 1989); Checkovich, "Mapping the American Way"; and Denise Phillips, "Friends of Nature: Urban Sociability and Regional Natural History in Dresden, 1800–1850," *Osiris* 18 (2003): 43–59.
51. Sharon Kingsland, *The Evolution of American Ecology, 1890–2000* (Baltimore: Johns Hopkins University Press, 2005), esp. chap. 9 and conclusion.
52. Grove, *Green Imperialism*; and Grove, "Indigenous Knowledge."
53. Emma C. Spary, *Utopia's Garden: French Natural History from Old Regime to Revolution* (Chicago: University of Chicago Press, 2000), chap. 2: Fa-ti Fan, "Victorian Naturalists in China: Science and Informal Empire," *British Journal for the History of Science* 36 (2003): 1–26; Fan, "Science in a Chinese Entrepôt: British Naturalists and Their Chinese Associates in Old Canton," *Osiris* 18 (2003): 60–78; and Fan, *British Naturalists in Qing China: Science, Empire, and Cultural Encounter* (Cambridge: Harvard University Press, 2004). See also Jim Endersby, *Imperial Nature: Joseph Hooker and the Practices of Victorian Science* (Chicago: University of Chicago Press, 2008).

54. Lisbet Koerner, *Linnaeus: Nature and Nation* (Cambridge: Harvard University Press, 1999); Staffan Müller-Wille, "Nature as a Marketplace: The Political Economy of Linnaean Botany," in Schabas and De Marchi, *Oeconomies in the Age of Newton*, 154–172; and Müller-Wille, "Joining Lapland."
55. Steven J. Harris, "Long-Distance Corporations, Big Sciences, and the Geography of Knowledge," *Configurations* 6 (1998): 269–304; and David S. Lux and Harold J. Cook, "Closed Circles or Open Networks? Communicating at a Distance During the Scientific Revolution," *History of Science* 36 (1998): 179–211.
56. For example, see Daniela Bleichmar, "Atlantic Competitions: Botany in the Eighteenth-Century Spanish Empire," in Delbourgo and Dew, *Science and Empire*, 225–252. This volume takes polycentrism as its central theme.
57. Alfred W. Crosby, *Ecological Imperialism: The Biological Expansion of Europe, 900–1900* (New York: Cambridge University Press, 1986); and Michael A. Osborne, "Acclimatizing the World: A History of the Paradigmatic Colonial Science," *Osiris* 15 (2000): 135–151. Osborne's paper is a good point of entry into a substantial literature on acclimatization science.
58. Janet Browne, *The Secular Ark: Studies in the History of Biogeography* (New Haven: Yale University Press, 1983); Browne, "A Science of Empire: British Biogeography Before Darwin," *Revue d'Histoire des Sciences* 45 (1992): 453–475; and Jane R. Camerini, "Evolution, Biogeography, and Maps: An Early History of Wallace's Line," *Isis* 84 (1993): 700–727.
59. An evocative literary meditation on human restlessness is Bruce Chatwin, *The Songlines* (New York: Viking, 1987).
60. Harold J. Cook, *Matters of Exchange: Commerce, Medicine, and Science in the Dutch Golden Age* (New Haven: Yale University Press, 2007); Cook, "The Cutting Edge of a Revolution: Medicine and Natural History near the Shores of the North Sea," in *Renaissance and Revolution: Humanists, Scholars, Craftsmen, and Natural Philosophers in Early Modern Europe*, ed. J. V. Field and Frank A. J. L. James (Cambridge: Cambridge University Press, 1993), 45–61; and Cook, "The Moral Economy of Natural History and Medicine in the Dutch Golden Age," in *Contemporary Explorations in the Culture of the Low Countries*, ed. William Z. Shetter and Inge Van der Cruysse (Lanham, Md.: University Press of America, 1996), 39–47. See also Cooper, *Inventing the Indigenous*; and Brian Ogilvie, *The Science of Describing: Natural History in Renaissance Europe* (Chicago: University of Chicago Press, 2006).
61. Antonio Barrera-Osorio, *Experiencing Nature: The Spanish American Empire and the Early Scientific Revolution* (Austin: University of Texas Press, 2006), 59–60; Barrera-Osorio, "Local Herbs, Global Medicines: Commerce, Knowledge, and Commodities in Spanish America," in *Merchants and Marvels: Commerce, Science, and Art in Early Modern Europe*, ed. Pamela H. Smith and Paula Findlen (New York: Routledge, 2002), 163–181; and Barrera Osorio, "Empiricism in the Spanish Atlantic World," in Delbourgo and Dew, *Science and Empire*, 177–202. Barrera-Osorio treats natural history as just one element of the imperial gathering of facts about navigation, geography, resources, peoples, and nature. Cook focuses on commodities and conditions of trade.
62. A useful point of entry into a substantial literature is Lynn K. Nyhart and Thomas H. Broman, eds., *Science and Civil Society, Osiris* 17 (2002). See especially Andreas

Daum, "Science, Politics, and Religion: Humboldtian Thinking and the Transformations of Civil Society in Germany, 1830–1870," *Osiris* 17 (2002): 107–140; and Elizabeth A. Hachten, "In Service to Science and Society: Scientists and the Public in Late-Nineteenth-Century Russia," *Osiris* 17 (2002): 171–209. See also Charles W. J. Withers, "Towards a History of Geography in the Public Sphere," *History of Science* 37 (1999): 45–78.

63. Diarmid A. Finnegan, "Natural History Societies in Late Victorian England and the Pursuit of Local Civic Society," *British Journal for the History of Science* 38 (2005): 53–72; Vladimir Jankovic, "The Place of Nature and the Nature of Place: The Chorographic Challenge to the History of British Provincial Science," *History of Science* 38 (2000): 79–113; Simon Naylor, "The Field, the Museum, and the Lecture Hall: The Spaces of Natural History in Victorian Cornwall," *Transactions of the Institute of British Geographers* 27 (2002): 494–513; Naylor, "Collecting Quoits: Field Cultures in the History of Cornish Antiquarianism," *Cultural Geographies* 10 (2003): 309–333; Charles Withers and Diarmid A. Finnegan, "Natural History Societies, Fieldwork, and Local Knowledge in Nineteenth-Century Scotland: Toward a Historical Geography of Civic Science," *Cultural Geographies* 10 (2003): 334–353; and Phillips, "Friends of Nature."
64. The best account of the "Heimat" movement is still Celia Applegate, *A Nation of Provincials: The German Idea of Heimat* (Berkeley and Los Angeles: University of California Press, 1990). Applegate is, however, more concerned with language and culture than sciences of place.
65. Lynn K. Nyhart, *Modern Nature: The Rise of the Biological Perspective in Germany* (Chicago: University of Chicago Press, 2009). See also Charles W. J. Withers, *Geography, Science, and National Identity: Scotland Since 1520* (Cambridge: Cambridge University Press, 2001); H. Glenn Penny, "Fashioning Local Identities in an Age of Nation-Building: Museums, Cosmopolitan Traditions, and Intra-German Competition," *German History* 17 (1999): 485–504; and Kapil Raj, "Colonial Encounters and the Forging of New Knowledge and National Identities: Great Britain and India, 1760–1850," *Osiris* 15 (2000): 119–134.
66. This idea is developed in Kohler, "Finders, Keepers."
67. For example, see Dorinda Outram, "New Spaces in Natural History," in Jardine, Secord, and Spary, *Cultures of Natural History*, 249–265 (on armchair versus field); Kuklick, "Personal Equations"; Emmanuelle Sibeud, "The Metamorphosis of Ethnology in France, 1839–1930," in Kuklick, ed., *A New History of Anthropology* (Oxford: Blackwell, 2008), 96–110; H. Glenn Penny, "Traditions in the German Language," in Kuklick, *New History*, 79–95; Kohler, *All Creatures*; Knell, *Culture of English Geology*; and Roy S. Porter, "Gentlemen and Geology: The Emergence of a Scientific Career, 1660–1920," *Historical Journal* 21 (1978): 809–836.
68. Bruce Hevly, "The Heroic Science of Glacier Motion," *Osiris* 11 (1996): 66–86; and Vetter and Rozwadowski essays, this volume.
69. For a general discussion, see Kuklick, "Personal Equations."
70. Brock, "Challenge of Reforestation." See also Matthias Gross, *Inventing Nature: Ecological Restoration by Public Experiments* (Lanham, Md.: Lexington Books, 2003).
71. Michael Bravo, "Mediating Field Station Histories: Voices from Igloolik" (unpublished conference MS, May 2007); J. Conor Burns, this volume; Jeremy Vetter,

"Cowboys, Scientists, and Fossils." As used by its inventor, Alan Irwin, the term "citizen science" meant something more like the older rubric of "science for the people": that is, expert science not for profit or pure thought, but for public good. Irwin, *Citizen Science: A Study of People, Expertise, and Sustainable Development* (London: Routledge, 1995).

72. Kohler, *All Creatures*, 156–162, 184–185.
73. Kohler, "Wildlife Ecology." Herbert Stoddard's research on quail is another such case. See Albert G. Way, "Stalking Wildlife Management: Predators, Prey, and the Development of a Profession" (unpublished MS, 2009); and Way, "Burned to be Wild."
74. George W. Stocking Jr., "The Ethnographer's Magic: Fieldwork in British Anthropology from Tylor to Malinowski," in *Observers Observed: Essays on Ethnographic Fieldwork*, ed. Stocking (Madison: University of Wisconsin Press, 1983), 70–120; Kuklick, "Personal Equations"; and Marcia Myers Bonta, *Women in the Field: America's Pioneering Women Naturalists* (College Park: Texas A&M University Press, 1991), chaps. 19–21. Recent histories of field ethology include: Richard W. Burkhardt Jr., *Patterns of Behavior: Konrad Lorenz, Niko Tinbergen, and the Founding of Ethology* (Chicago: University of Chicago Press, 2005); Burkhardt, "Ethology, Natural History, the Life Sciences, and the Problem of Place," *Journal of the History of Biology* 32 (1999): 489–508; Amanda Rees, "A Place That Answers Questions: Primatological Field Sites and the Making of Authoritative Observations," *Studies in History and Philosophy of Biological and Biomedical Sciences* 37 (2006): 311–333; *Rees, The Infanticide Controversy: Primatology and the Art of Field Science* (Chicago: University at Chicago Press 2009); and Georgina M. Montgomery, "Place, Practice, and Primatology: Clarence Ray Carpenter, Primate Communication, and the Development of Field Methodology, 1931–1945," *Journal of the History of Biology* 38 (2005): 495–533.
75. Dale Peterson, *Jane Goodall: The Woman Who Redefined Man* (Boston: Houghton Mifflin, 2006); and Jonathan Weiner, *The Beak of the Finch: A Story of Evolution in Our Time* (New York: Knopf, 1994). A study of settlement-house science is needed.

Notes on Contributors

Mark V. Barrow Jr. is a professor in the History Department and an affiliated faculty member in the Science and Technology in Society Department at Virginia Tech. His research centers on the intersection between the history of natural history, wildlife conservation, and American culture. His first book, *A Passion for Birds: American Ornithology After Audubon* (Princeton University Press, 1988), won the Forum for the History of Science in America Book Prize, while his second book, *Nature's Ghosts: Confronting Extinction from the Age of Jefferson to the Age of Ecology*, was published by the University of Chicago Press in 2009. His is currently working on a cultural and environmental history of the American alligator.

J. Conor Burns currently teaches history of science and technology at York University and Ryerson University, both in Toronto, and is Assistant Book Review Editor for Isis. His work broadly addresses relationships between the historical, human, and field sciences, and his current research examines how an emerging science of archaeology shaped conceptualizations of the past in nineteenth- and twentieth-century North America. He has recently published "Networking Ohio Valley Archaeology in the 1880s: The Social Dynamics of Peabody and Smithsonian Centralization," in *Histories of Anthropology Annual* 4 (2008): 1–33.

James Rodger Fleming is Professor of Science, Technology, and Society at Colby College. His books include *Fixing the Sky: The Checkered History of Weather and Climate Control* (Columbia University Press, 2010), *The Callendar*

Effect (American Meteorological Society, 2007), *Historical Perspectives on Climate Change* (Oxford University Press, 1998), and *Meteorology in America, 1800–1870* (Johns Hopkins University Press, 1990). Recent co-edited volumes include *Intimate Universality* (Science History/USA, 2006) and *Osiris* 26, *Revisiting Klima* (forthcoming). He is working on connecting the local and global in the history of Earth system science and on connecting the history of science and technology with public policy.

Robert E. Kohler is Professor Emeritus in the Department of History and Sociology of Science at the University of Pennsylvania. He is the author of numerous articles and books, including *All Creatures: Naturalists, Collectors, and Biodiversity, 1850–1950* (Princeton University Press, 2006), *Landscapes and Labscapes: Exploring the Lab-Field Border in Biology* (University of Chicago Press, 2002), and *Lords of the Fly: Drosophila Genetics and the Experimental Life* (University of Chicago Press, 1994).

Stuart McCook is Associate Professor of History at the University of Guelph. He is author of *States of Nature: Science, Agriculture, and Environment in the Spanish Caribbean, 1760–1940* (University of Texas Press, 2002) and has published articles in several journals, including *The Americas, Journal of Global History, Agricultural History, Endeavour*, and *Osiris*. His research focuses on the history of the agricultural sciences, as well as the environmental history of agriculture in the global tropics. He is currently writing a global history of the coffee rust, an epidemic disease of the coffee plant, as a way of exploring environmental processes linking the coffee industries of Asia, Africa, Latin America, and the Pacific.

Lynn K. Nyhart is Professor in the Department of the History of Science at the University of Wisconsin, Madison, where she specializes in history of modern biology. The author of *Biology Takes Form: Animal Morphology and the German Universities* (University of Chicago Press, 1995), she has recently completed a book on the prehistory of ecology in Germany, *Modern Nature: The Rise of the Biological Perspective in Germany* (University of Chicago Press, 2009). She is currently pursuing two projects, on the history of the aquarium and on the history of ideas about parts and wholes in biology.

Michael S. Reidy is Associate Professor and Director of Graduate Studies in the Department of History and Philosophy at Montana State University. He is coauthor of *Communicating Science: The Scientific Journal Article from the Seventeenth Century to the Present* (Oxford University Press, 2000) and author of *Tides of History: Ocean Science and Her Majesty's Navy* (University of

Chicago Press, 2008). He is currently working on a history of British mountaineering and science in the nineteenth century.

Helen M. Rozwadowski is Associate Professor and Maritime Studies Coordinator at the University of Connecticut, Avery Point. While working as the historian for the International Council for the Exploration of the Sea she wrote *The Sea Knows No Boundaries: A Century of Marine Science Under ICES* (University of Washington Press, 2002). Her most recent book, *Fathoming the Ocean: The Discovery and Exploration of the Deep Sea* (Harvard University Press, 2005), is a scientific and cultural history of interest in the ocean, manifested in maritime novels, the popular hobby of marine zoology, the youthful sport of yachting, and the laying of the transatlantic telegraph cable. Her current research again examines the confluence of scientific and popular interest in the ocean, this time during the cold war, at a time when Western nations conceived of the sea as a new frontier akin to outer space.

Jeremy Vetter is Assistant Professor of History at the University of Arizona in Tucson. He has published articles in the *Journal of the History of Biology, Annals of Science, Notes and Records of the Royal Society*, and *Isis*. His is currently working on two book manuscripts tentatively titled “Field Life: Doing Science in the American West During the Railroad Era” and “Capitalist Nature: Environment, Science, and Development in the U.S. Great Plains and Rocky Mountains, 1860–1920.”

Index